Two Faces of Evil: Cancer and Neurodegeneration

For further volumes:
http://www.springer.com/series/1175

Thomas Curran · Yves Christen

Editors

Two Faces of Evil: Cancer and Neurodegeneration

Springer

Editors
Thomas Curran, Ph.D., FRS
The Children's Hospital of Philadelphia
Department of Pathology
and Laboratory Medicine
Civic Center Boulevard 3501
Philadelphia, PA 19104
Pennsylvania
USA
currant@email.chop.edu

Yves Christen, Ph.D
Fondation IPSEN pour la
Recherche Therapeutique
65 quai Georges Gorse
92650 Boulogne-Billancourt
Cedex
France
yves.christen@beaufour-ipsen.com

ISSN 0945-6066
ISBN 978-3-642-16601-3   e-ISBN 978-3-642-16602-0
DOI 10.1007/978-3-642-16602-0
Springer Heidelberg Dordrecht London New York

Cover design: WMXDesign GmbH, Heidelberg, Germany

Printed on acid-free paper

Springer is part of Springer Science+Business Media (www.springer.com)

Foreword

Homeostasis involves a delicate interplay between generative and degenerative processes to maintain a stable internal environment. In biological systems, equilibrium is established and controlled through a series of negative feedback mechanisms driven by a range of signal transduction processes. Failures in these complex communication pathways result in instability leading to disease. Cancer represents a state of imbalance caused by an excess of cell proliferation. In contrast, neurodegeneration is a consequence of excessive cell loss in the nervous system. Both of these disorders exhort profound tolls on humanity and they have been subject to a great deal of research designed to ameliorate this suffering. For the most part, the topics have been viewed as distinct and rarely do opportunities arise for transdisciplinary discussions among experts in both fields. However, cancer and neurodegeneration represent *yin–yang* counterpoints in the regulation of cell growth, and it is reasonable to hypothesize that key regulatory events mediated by oncogenes and tumor suppressor genes in cancer may also affect neurodegenerative processes. This was the rationale for organizing the *Colloques Médecine et Recherche*, April 26, 2010 on the topic of *Two Faces of Evil: Cancer and Neurodegeneration*.

The presentations by the leaders of both fields were full of exciting unpublished data that reaffirmed the connection between the disciplines. Remarkably, genes that affect cell cycle progression and checkpoint control also play specific roles in postmitotic neurons that influence neurodegeneration. Many oncogenic signaling pathways have been expropriated to fulfill distinct functions in a range of biological situations. The complexity of the nervous system is such that evolution has usurped most molecular, biochemical and cellular regulatory mechanisms to support the formation, function and maintenance of neurons. The discussions at the meeting transcended traditional boundaries. Several new concepts were shared that will stimulate future research and perhaps contribute to better therapies for cancer patients as well as those struggling with the ravages of neurodegeneration.

Acknowledgments

The editors would like to extend profound thanks to Jacqueline Mervaillie and Sonia Le Cornec for their smooth organization of the meeting, despite *Eyjafjallajökull*, and to Astrid de Gérard for her efficient, gentle persistence in putting this book together.

Contents

Contributors

Arturo Alvarez-Buylla Department of Neurological Surgery and The Eli and Edythe Broad Center of Regeneration Medicine and Stem Cell Research, University of California, San Francisco, CA 94143, USA, abuylla@stemcell.ucsf.edu

Cristine Alves da Costa IPMC and IN2M, UMR6097 CNRS/UNSA, Team Fondation pour la Recherche Médicale, Sophia-Antipolis, 06560, Valbonne, France

Mariano Barbacid Centro Nacional de Investigaciones Oncológicas (CNIO), Melchor Fernández Almagro 3, 28029 Madrid, Spain, barbacid@cnio.es

Sourav Banerjee Neuroscience Research Institute, Department Molecular Cellular Developmental Biology, University of California, Santa Barbara, CA 93106, USA

Frédéric Checler IPMC and IN2M, UMR6097 CNRS/UNSA, Team Fondation pour la Recherche Médicale, Sophia-Antipolis, 06560 Valbonne, France, checler@ipmc.cnrs.fr

Tom Curran Department of Pathology and Laboratory Medicine, The Children's Hospital of Philadelphia, Philadelphia, PA 19104, USA, currant@email.chop.edu

Robert B. Darnell Laboratory of Molecular Neuro-Oncology, The Rockefeller University, Box 226, 1230 York Avenue, New York, NY 10021, USA, darnelr@rockefeller.edu

Julie Dunys IPMC and IN2M, UMR6097 CNRS/UNSA, Team Fondation pour la Recherche Médicale, Sophia-Antipolis, 06560 Valbonne, France

Michael E. Greenberg Department of Neurobiology, Harvard Medical School, 220 Longwood Avenue, Boston, MA 02115, USA, meg@hms.harvard.edu

Paul L. Greer Department of Neurobiology, Harvard Medical School, 220 Longwood Avenue, Boston, MA 02115, USA

Eric C. Griffith Department of Neurobiology, Harvard Medical School, 220 Longwood Avenue, Boston, MA 02115, USA

Junjie U. Guo Institute for Cell Engineering, Department of Neuroscience, Johns Hopkins University School of Medicine, 733 N. Broadway, BRB 759, Baltimore, MD 21205, USA

Young-Goo Han Department of Neurological Surgery and The Eli and Edythe Broad Center of Regeneration Medicine and Stem Cell Research, University of California, San Francisco, CA 94143, USA

Karl Herrup Department of Cell Biology and Neuroscience, Rutgers University, 604 Allison Road, Piscataway, NJ 08854, USA, herrup@biology.rutgers.edu

Eric Kandel Department of Neuroscience, Howard Hughes Medical Institute, Kavli Institute for Brain Sciences, Columbia University, New York, NY 10032, USA; The Scripps Research Institute, Scripps Florida, 130 Scripps Way, Jupiter, FL 33458, USA, erk5@columbia.edu

David R. Kaplan Department of Molecular Genetics, Cell Biology Programs, Hospital for Sick Children, University of Toronto, Toronto, ON, Canada M5G 1L7

Bernd Knöll Neuronal Gene Expression Laboratory, Department of Molecular Biology, Interfaculty Institute for Cell Biology, University of Tübingen, Auf der Morgenstelle 15, 72076 Tübingen, Germany

Kenneth S. Kosik Neuroscience Research Institute, Department Molecular Cellular Developmental Biology, University of California, Santa Barbara, CA 93106, USA, kosik@lifesci.ucsb.edu

Dengke K. Ma Institute for Cell Engineering, Department of Neuroscience, Johns Hopkins University School of Medicine, 733 N. Broadway, BRB 759, Baltimore, MD 21205, USA; Department of Biology, Massachusetts Institute of Technology, 77 Massachusetts Ave., Room 68-441, Cambridge, MA 02139, USA

Freda D. Miller Developmental and Stem Cell Biology and Departments of Molecular Genetics and Physiology, University of Toronto, Toronto, ON Canada M5G 1L7

Guo-li Ming Institute for Cell Engineering, Department of Neuroscience, Department of Neurology, Johns Hopkins University School of Medicine, 733 N. Broadway, BRB 759, Baltimore, MD 21205, USA

Pierre Neveu Neuroscience Research Institute, Department Molecular Cellular Developmental Biology, University of California, Santa Barbara, CA 93106, USA

Alfred Nordheim Vertebrate Gene Expression and Organ Function, Department of Molecular Biology, Interfaculty Institute for Cell Biology, University of Tübingen, Auf der Morgenstelle 15, 72076 Tübingen, Germany, alfred.nordheim@uni-tuebingen.de

Harry T. Orr Institute of Translational Neuroscience, Department of Laboratory Medicine and Pathology, University of Minnesota, Minneapolis, MN 55455, USA, orrxx002@umn.edu

Raphaëlle Pardossi-Piquard IPMC and IN2M, UMR6097 CNRS/UNSA, Team Fondation pour la Recherche Médicale, Sophia-Antipolis, 06560, Valbonne, France

Sathyanarayanan Puthanveettil Department of Neuroscience, Howard Hughes Medical Institute, Chevy Chase, MD USA; Department of Neuroscience, The Scripps Research Institute, Scripps Florida, 130 Scripps Way, Jupiter, FL 33458, USA, sp2068@columbia.edu

Hongjun Song Institute for Cell Engineering, Department of Neuroscience, Department of Neurology, Johns Hopkins University School of Medicine, 733 N. Broadway, BRB 759, Baltimore, MD 21205, USA, shongju1@jhmi.edu

Updating the Mammalian Cell Cycle: The Role of Interphase Cdks in Tissue Homeostasis and Cancer

Mariano Barbacid

Abstract Genetic interrogation of the mammalian cell cycle has revealed that the essential role of interphase Cdks is not to specifically drive the various phases of the cycle, as previously proposed in widely accepted models, but to sustain proliferation of specialized cells at various times during embryonic or postnatal development. Indeed, genetic studies have indicated that Cdk1 can drive the mammalian cell cycle in the absence of interphase Cdks. The molecular bases for the essential requirement of interphase Cdks in selected cell types are still poorly understood. However, these observations have important implications for understanding the role of Cdk misregulation in cancer. Indeed, it is likely that misregulation of Cdks may only confer proliferative advantages to selected cell types. More importantly, it is also possible that certain cells may become dependent of selective interphase Cdks only when their proliferation is driven by defined oncogenes. Recent studies have illustrated the requirement for Cdk4 in HER2-overexpressed mammary adenocarcinomas and K-Ras oncogene-driven lung adenocarcinomas but not in the corresponding normal tissues. Likewise, Cdk2 plays an important role in the development of Myc-induced lymphomas. These findings may open the door to the design of novel therapeutic strategies that may benefit cancer patients.

Molecular analysis of human tumors has revealed that most carry mutations that result in misregulation of the cell cycle. Tumor cells accumulate mutations that result in unscheduled cell proliferation, either by constitutive activation of mitogenic signaling pathways or by elimination of regulatory anti-mitogenic signals (Malumbres and Barbacid 2001; Massague 2004). In addition, most tumors acquire genomic instability as a consequence of alterations in cell cycle checkpoints (Kastan and Bartek 2004) and chromosome instability due to errors during the mitotic phase of the cycle (Kops et al. 2005; Fig. 1). Progression through the cell cycle is driven by heterodimeric kinases made up of a regulatory subunit, known as

M. Barbacid
Molecular Oncology Programme, Centro Nacional de Investigaciones Oncológicas (CNIO), Melchor Fernández Almagro 3, E-28029 Madrid, Spain
e-mail: barbacid@cnio.es

T. Curran and Y. Christen (eds.), *Two Faces of Evil: Cancer and Neurodegeneration*,
Research and Perspectives in Alzheimer's Disease, DOI 10.1007/978-3-642-16602-0_1,
© Springer-Verlag Berlin Heidelberg 2011

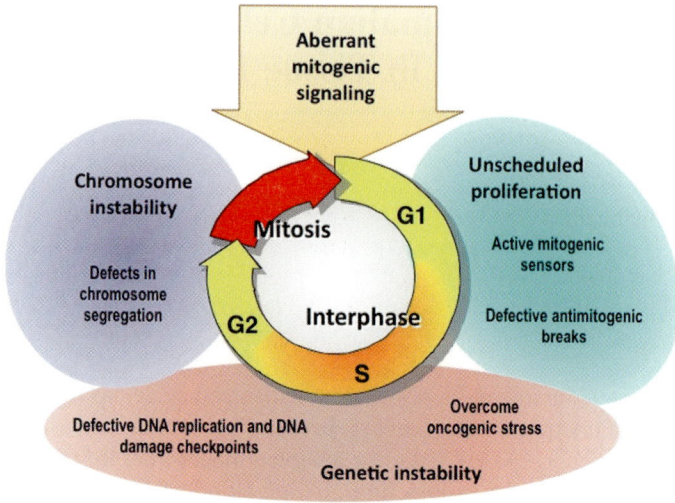

Fig. 1 Schematic diagram of the mammalian cell cycle with those defects implicated in human cancer

Cyclin, and a catalytic subunit, designated as Cyclin-dependent kinase or Cdk. This designation stems from the fact that this catalytic subunit is completely inactive in the absence of its corresponding activating subunit, the Cyclin. Cdks are constitutively expressed during the cycle. In contrast, Cyclins are synthesized and destroyed at specific phases, thereby regulating kinase activity in a timely manner during the cell cycle (Malumbres and Barbacid 2001).

Mammalian cells express multiple Cyclin and Cdks, although not all of them are involved in driving the cell cycle (Malumbres and Barbacid 2005). Among the Cdks, only five, Cdk1, Cdk2, Cdk3, Cdk4 and Cdk6, are thought to directly participate in cell cycle regulation. Cdk7 also plays an important role in the cell cycle since this kinase is responsible for activating Cdk1 and possibly Cdk2. Whereas Cdk1 is generally considered to be "the" mitotic kinase, the other Cdks are believed to play specific roles in the distinct phases of cell division that constitute the interphase, that is, the period between two mitotic events in proliferating cells (see below). Among these kinases, only Cdk4 has been found to be mutated in human cancer, and only in a few cases of hereditary melanoma (Wölfel et al. 1995). In addition, Cdk6 overexpression has been documented in lymphomas, leukemias and melanomas as a consequence of chromosomal translocations. However, Cdk activity can be altered in human cancer by several independent mechanisms, including increased levels of Cyclin expression or impaired degradation that result in increased availability of these regulatory subunits. Increased Cdk activity can also result from loss of members of the INK4 and CIP/KIP families of Cdk inhibitors. Whereas INK4 proteins bind directly to the Cdk catalytic subunits, preventing their interaction with their cognate Cyclins, the CIP/KIP inhibitors bind to Cdk/Cyclin heterodimers, forming inactive tertiary complexes (Malumbres and Barbacid 2001; Massague 2004).

Little is known about the full spectrum of physiological substrates of the interphase Cdk/Cyclin complexes. However, the role of these kinases in inactivating the pocket proteins – RB, p107 and p130 – has been profusely illustrated (Classon and Harlow 2002). These proteins function as constitutive repressors of the cell cycle machinery in quiescent cells. Tumor-associated alterations frequently deregulate these Cyclin–Cdk complexes, resulting in constitutive inactivation of the pocket proteins and allowing either continued proliferation or unscheduled re-entry into the cell cycle, two properties characteristic of almost every human cancer cell (Malumbres and Barbacid 2001).

1 Mammalian Cdks and the Classical Cell Cycle Model

The basic regulation of the cell cycle by Cdk/Cyclin complexes was first established in yeasts. In these unicellular organisms, cell cycle progression is driven by a single Cdk – Cdc28 and Cdc2, the orthologs of mammalian Cdk1 – that binds sequentially to various Cyclins at different stages of the cycle (Nurse 1997). In mammals, the number of Cdks and Cyclins implicated in the control of cell cycle progression has increased considerably (Malumbres and Barbacid 2005). According to a widely accepted model for the mammalian cell cycle, these additional Cdks, generated throughout evolution, have acquired unique specific roles during interphase by overtaking the roles played by Cdc28/Cdc2 during interphase and limiting Cdk1 activity to the mitotic phase (Fig. 2). This model has been primarily derived from biochemical evidence mostly gathered from studies carried out with human tumor cells grown in culture for many generations. Unfortunately, recent genetic evidence has proven to be incompatible with some key tenants proposed by the classical model.

According to this model, Cdk/Cyclin complexes containing the mammalian interphase Cdks – Cdk2, Cdk3, Cdk4 and Cdk6 – (Cdk5 plays a specific role in the brain and will be discussed in other chapters) bound to their cognate Cyclins act in a sequential and orderly fashion during the various events known to take place during interphase. That is, this model replaces the basic yeast cell cycle model in that the kinases responsible for driving the cell cycle are made up of different catalytic and regulatory subunits instead of a single catalytic subunit (Cdc28/Cdc2) sequentially bound to the different Cyclins (Fig. 2). According to this model, mitogenic signals are first sensed by inducing expression of D-type Cyclins (D1, D2 and D3) that bind and activate Cdk4 and Cdk6 during G1, a phase of the cell cycle in which cells prepare to initiate DNA synthesis. Recent evidence obtained in our laboratory has indicated that this concept may have to be revised, since non-proliferating cells with impaired mitogenic signaling due to ablation of the three Ras proteins express even higher than normal levels of Cyclin D1 (Drosten et al. 2010). Regardless of the mechanism by which D-type Cyclins are induced, they bind to and activate Cdk4 and/or Cdk6, which in turn

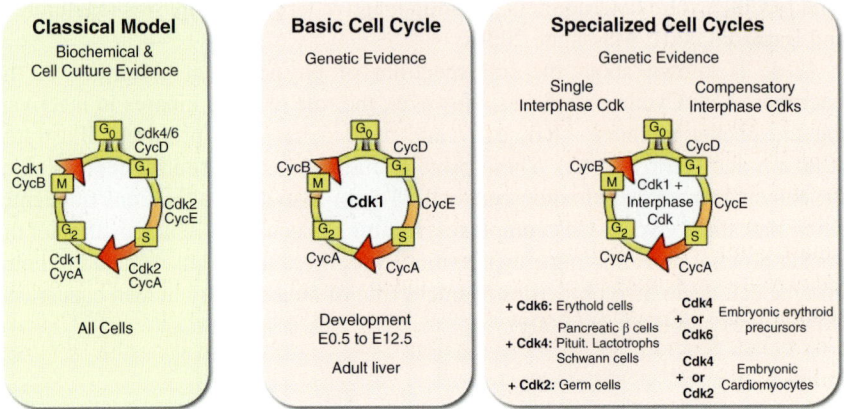

Fig. 2 Genetic interrogation of the roles of Cdks in the mammalian cell cycle. (*Left*) Classical model for the mammalian cell cycle deduced from biochemical and cell culture studies (reviewed in Malumbres and Barbacid 2001). (*Center*) Model for the basic mammalian cell cycle based on genetic evidence derived from ablating all interphase Cdks in the mouse embryo (Santamaria et al. 2007). A single Cdk, Cdk1, can drive the mammalian cell cycle by sequentially binding to the various Cyclins in a fashion similar to that proposed for the yeast cell cycle. (*Right*) Schematic representation of the involvement of the interphase Cdks in driving the cell cycle of specialized cell types (Specialized Cell Cycles). The types of cells requiring a single interphase Cdk are indicated at the left of the diagram. Those cell types requiring one of two different interphase Cdks are indicated at the right. It is assumed that all of these specialized cell cycles also requires Cdk1 since, without this enzyme, embryonic development does not progress beyond the two-cell stage. Yet, information regarding the need for Cdk1 expression in selective cell types is not available

phosphorylate and partially inactivate the pocket proteins. This limited inactivation is believed to allow expression of a subset of genes required for progression of the cell cycle into the S phase. A central role in this process is played by the E-type Cyclins (E1 and E2), which bind and activate Cdk2. Availability of E-type Cyclins during the cell cycle is a tightly controlled process limited to the early stages of DNA synthesis (S phase). Once the E-type Cyclins are degraded, Cdk2 binds to Cyclin A2 (to Cyclin A1 in germ cells). Considering that availability of A-type Cyclins coincides with the late stages of DNA replication, it has been postulated that Cdk2 exchanges partners during this critical phase of the cycle, to ensure proper completion of DNA synthesis and orderly transition from S phase to G2. Cdk1, the ortholog of the single yeast cell cycle Cdk, is thought to play a role in late G2 to facilitate entry into M phase. At this time, Cdk1 is activated by the same A-type Cyclins previously bound to Cdk2. Since Cdk2 is not degraded during G2, it is not clear how or why the A-type Cyclins switch partners, although it has been proposed that Cdk1/CyclinA complexes have more robust kinase activity, which is necessary to initiate mitosis. Upon nuclear envelop breakdown, A-type Cyclins are degraded, allowing Cdk1 to bind the B-type Cyclins, mainly Cyclin B1, to form the kinase complex responsible for driving cells through mitosis (Malumbres and Barbacid 2001).

2 Interphase Cdks are Not Essential for the Mammalian Cell Cycle

This classical model has been seriously challenged by genetic evidence derived from the analysis of mice defective for each of the interphase Cdks (Table 1). First of all, Ye et al. (2001) demonstrated that Cdk3 is not required, at least for laboratory mice, since most strains carry a homozygous nonsense mutation in the middle of the kinase domain that eliminates Cdk3 activity. Thus, ablation of any other interphase Cdks results in double or multiple Cdk mutant strains, since all the genetic manipulations published thus far have been done on genetic backgrounds deficient for Cdk3. Yet, for simplicity, this fact will not be considered for the work discussed below (Table 1).

Ablation of Cdk4 results in developmental defects derived from limited proliferation of highly specialized cells, but there is no evidence of an overall defect in cell cycle entry or in progression through the G1 phase in most cells (see below). The main cell types affected by loss of Cdk4 expression include insulin-producing pancreatic beta cells and pituitary endocrine cells (Fig. 3; Rane et al. 1999; Tsutsui et al. 1999; Moons et al. 2002). Interestingly, Cdk4 is only required during postnatal development of these cell types. More recently, Atanasoski et al. (2008) have shown that Cdk4, but not Cdk2 or Cdk6, is also essential for postnatal proliferation of Schwann cells. Longitudinal sections of P2 Cdk4 *null* nerves double-stained with

Table 1 Genetic ablation of mammalian Cdks in mice: phenotypic consequences

Kinase	Genotype[a]	Phenotype	References
Cdk3	$Cdk3^{Stop/Stop}$	Natural mutation in certain strains of inbred mice. No detectable phenotype.	Ye et al. (2001)
Cdk4	$Cdk4^{-/-}$	Defects in postnatal proliferation of highly specialized cells, including insulin-producing pancreatic beta cells, hormone-producing pituitary lactotrophs and Schwann cells. As a consequence, these mutant mice develop diabetes and display defective Schwann cell proliferation upon nerve injury. Females are sterile.	Rane et al. (1999), Tsutsui et al. (1999), Moons et al. (2002), Atanasoski et al. (2008)
Cdk6	$Cdk6^{-/-}$	Mice display minor hematopoietic defects, including slight anemia, due to reduced number of cells of erythroid lineage and a defective proliferative response of T lymphocytes upon mitogenic stimuli.	Malumbres et al. (2004)
Cdk2	$Cdk2^{-/-}$	Mice are normal except for male and female sterility due to defective meiosis. No defects in mitotic cell division.	Berthet et al. (2003), Ortega et al. (2003)
Cdk1	$Cdk1^{mut/mut}$	Embryonic lethality due to lack of cell division beyond the two-cell embryo stage.	Santamaría et al. 2007

[a]$Cdk4^{-/-}$, $Cdk6^{-/-}$, $Cdk2^{-/-}$ and $Cdk2^{mut/mut}$ strains also carry the $Cdk3^{Stop/Stop}$ germ line mutation (Ye et al. 2001) and hence are defective for Cdk3

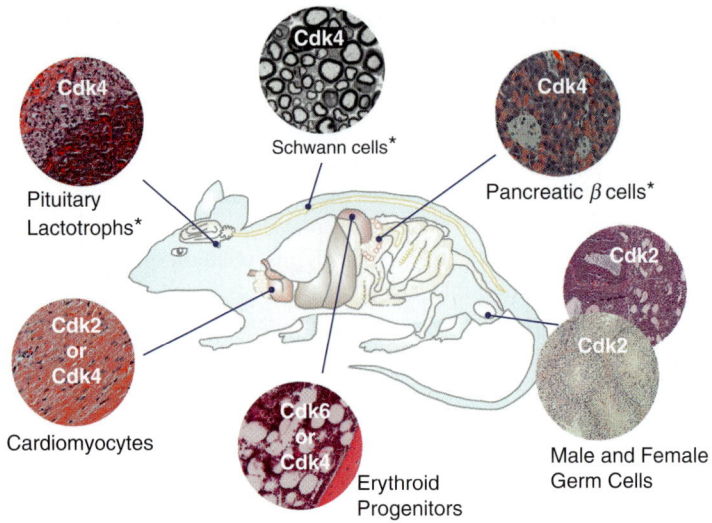

Fig. 3 Diagram depicting the specific requirement for interphase Cdks in specialized cell types. The indicated Cdks are essential for the corresponding cell type (e.g. Cdk4 is essential for proliferation of pancreatic β cells). In those cases in which two Cdks are listed, the indicated cells proliferate well in the absence of one Cdk but not in the absence of both Cdks, due to compensatory activities. The asterisk indicates that Cdk4 is only required during postnatal development

the Ki-67 and DAPI displayed a striking decrease in the number of mitotically active Schwann cells when compared with those of wild type mice. No such differences were observed in mice lacking Cdk2 or Cdk6 (Atanasoski et al. 2008). Proliferation rates of Cdk4-deficient Schwann cells were approximately 20-fold lower than those of wild type cells at both P2 and P5. These observations were not due to increased apoptosis. They illustrate another essential role of Cdk4 in highly specialized postnatal cells (Table 1). Interestingly, loss of Cdk4 had no effect on the myelination of the sciatic nerve or on the length of myelinated internotes (Atanasoski et al. 2008). Thus, postnatal proliferation is not required for the establishment of appropriate Schwann cell numbers, suggesting that the number of Schwann cells produced before birth is sufficient to ensure subsequent proper myelination of axons. Atanasoski et al. (2008) also studied whether Cdk4 expression was required for proliferation of adult Schwann cells in Wallerian degeneration. Whereas no proliferating Schwann cells could be seen in unaffected normal nerves, there were numerous proliferating cells in lesioned nerves four days after injury. However, no dividing Schwann cells could be detected in the injured mice lacking Cdk4 (Atanasoski et al. 2008). Thus, adult Schwann cells must require Cdk4 expression to reenter the cell cycle upon nerve axotomy. Interestingly, when nerve injury was performed in mice lacking Cdk2 or Cdk6, no differences were observed between wild type and mutant mice (Atanasoski et al. 2008). These recent findings further illustrate that the requirements for Cdk4 are not cell-cycle dependent but cell-type dependent.

Mice lacking Cdk6 in their germ line are born according to Mendelian ratios, develop normally and are fertile, thus indicating that Cdk6 is not essential for the mammalian cell cycle (Malumbres et al. 2004). Detailed examination of the hematopoietic system in adult mice, a tissue in which Cdk6 is primarily expressed, revealed minor but reproducible defects. For instance, both the thymus and spleen were reduced in size due to lower cellularity. In the latter, the defect was exclusively observed in the red pulp, a defect due to a decrease in the number of erythroid cells (Malumbres et al. 2004). The number of megakaryocytes was also reduced in these mutant mice. Red cells in peripheral blood were also decreased by about 15%. The numbers of granulocytes, macrophages, neutrophils, and platelets were also reduced in the peripheral blood of some, but not all, mutant mice. B and T cells were present at the expected ratios. In the thymus, there was a slight but consistent increase of CD4 and CD8 single positive populations and a consequent decrease of immature double-positive cells. Finally, T lymphocytes lacking Cdk6 displayed a delayed response to mitogenic stimuli (Malumbres et al. 2004). Yet, none of the other cells in these targeted mice displayed specific cell cycle defects that affected either re-entry into the cycle or passage through the G1 phase, two of the specific activities attributed to the Cdk6 kinase by the classical model.

Finally, Cdk2, the enzyme proposed to be essential to drive DNA synthesis according to the classical model, is also dispensable for the mammalian cell cycle (Berthet et al. 2003; Ortega et al. 2003; Table 1). Mice lacking Cdk2 in their germ line are completely normal except for their germ cells (Fig. 3). A defect in the first meiotic cycle results in complete elimination of germ cells from both male and female Cdk2 *null* animals, resulting in complete sterility (Ortega et al. 2003). Yet, Cdk2 appears to be completely dispensable for mitotic cell division.

3 Limited Compensatory Activities Among Interphase Cdks

The absence of basic cell cycle defects in mice lacking individual interphase Cdks is not due to compensatory activities among these kinases (Table 2). Indeed, concomitant ablation of two or three interphase Cdks – always in a genetic background already defective for Cdk3 – did not lead to significant defects in the cell cycle, except for in highly specialized cells. The most significant compensatory activity was observed between Cdk4 and Cdk6, two catalytic subunits known to have indistinguishable affinities for the D-type Cyclins as well as similar substrates. Unlike mice lacking Cdk4 or Cdk6 alone, mice lacking Cdk4 and Cdk6 die during late embryonic development. However, the embryonic lethality displayed by these mice was not due to an overall defect in the cell cycle in all or most embryonic cells but to limited proliferation of erythroid precursors. No specific defects in G1 phase or cell cycle re-entry were observed either. Thus, Cdk4/Cdk6 *null* mice die due to a

Table 2 Limited compensatory activities of mammalian interphase Cdks in mice: genetic evidence

Kinases[a]	Genotype[a,b]	Phenotype	References
Cdk2 and Cdk6	$Cdk2^{-/-}$; $Cdk6^{-/-}$	Same phenotypes as in single mutant mice. The lack of additional or exacerbated phenotypes indicates lack of compensatory effects between these kinases.	Malumbres et al. (2004)
Cdk4 and Cdk6	$Cdk4^{-/-}$; $Cdk6^{-/-}$	Embryos die at late gestation due to defects in proliferation of erythroid precursors. No cell cycle defects were observed in other embryonic tissues, indicating that the compensatory effects between these kinases are limited to this cell type.	Malumbres et al. (2004)
Cdk2 and Cdk4	$Cdk2^{-/-}$; $Cdk4^{-/-}$	Mice complete embryonic development but die right after birth, most likely due to limited number of cardiomyocytes. These observations indicate that Cdk2 and Cdk4 only have compensatory activities in this specialized cell type or in its progenitors.	Barrière et al. 2007
Cdk2 and Cdk4	$Cdk2^{lox/lox}$; $Cdk4^{-/-}$; $RERT^{ert/ert}$	Adult mice lacking Cdk2 and Cdk4 display the same phenotypes as the single mutant animals, indicating lack of compensatory effects between these kinases in adult mice. These double-mutant mice also display normal liver regeneration upon partial hepatectomy.	Barrière et al. 2007
Cdk2, Cdk4 and Cdk6	$Cdk2^{-/-}$; $Cdk4^{-/-}$; $Cdk6^{-/-}$	Embryos lacking all interphase Cdks develop normally up to midgestation. They fail to progress due to hematopoietic defects similar to those observed in $Cdk4^{-/-}$;$Cdk6^{-/-}$ embryos. Loss of Cdk2 slightly exacerbates this phenotype. No additional defects compared with these mice were observed, indicating that Cdk1 was capable of driving the cell cycle in all other cell types.	Santamaría et al. 2007

[a]In addition to the indicated mutations, all mice also carry the $Cdk3^{Stop/Stop}$ germ line mutation (Ye et al. 2001) and hence are defective for Cdk3
[b]The $RERT^{ert}$ allele corresponds to the allele encoding the large subunit of RNA polymerase II, in which we have knocked-in an inducible CreERT2 recombinase using a bicistronic strategy (Guerra et al. 2003)

developmental defect and not to a basic cell cycle deficiency (Malumbres et al. 2004). Similar results have been observed when the three D-type Cyclins were ablated (Kozar et al. 2004), thus indicating that Cdk4/6-D type Cyclin kinase activity is not essential either to drive cells through G1 in particular or to drive the mammalian cell cycle in general, a concept that does not support one of the basic tenants of the classical model for the mammalian cell cycle (Fig. 2).

Mice lacking Cdk2 and Cdk6 reached adulthood and did not display any defects except those observed in the single mutant strains (Table 2; Malumbres et al. 2004). Likewise, mice lacking Cdk2 and Cdk4 completed embryonic development and did not show cell cycle defects except in embryonic cardiomyocytes, another highly specialized cell type (Barrière et al. 2007). These observations were not unique to embryonic cells, since adult mice lacking Cdk2 and Cdk4 in most of their cells do not display obvious defects. Indeed, these mice recovered normally after severe hepatectomy, indicating that adult hepatocytes proliferate normally without Cdk2 and Cdk4 kinases (Table 2; Barrière et al. 2007).

The most striking result derived from genetic interrogation of the role of interphase Cdks in the mammalian cell cycle came from the observation that mouse embryos developed normally until mid gestation without interphase Cdks (Santamaría et al. 2007; Fig. 2). Early development is the period of most active cell division and the time when organogenesis takes place. Yet, Cdk1 appears to be capable of sustaining millions of cell divisions, including normal organogenesis, in the absence of all interphase Cdks. Indeed, biochemical evidence indicates that, in the absence of interphase Cdks, Cdk1 can bind to all the cell cycle Cyclins, including D-type and E-type as well as its well-known partners, the A-type and B-type Cyclins (Aleem et al. 2005; Santamaría et al. 2007). These observations indicate that Cdk1 can sustain the cell cycle of all cell types by sequentially interacting with all type of Cyclins (Fig. 2). In cultured fibroblasts, the absence of interphase Cdks did not impede cell division, although fibroblasts expressing only Cdk1 display decreased proliferative potential, probably due to the load of cell cycle inhibitors expressed under these stressed culture conditions (Santamaría et al. 2007). Yet, even under these adverse conditions, Cdk1 drove these fibroblasts to immortality upon continuous passage. These observations, taken together, indicate that the mammalian cell cycle is not substantially different from that of yeasts and that the generation of interphase Cdks during evolution was not due to a need to drive selected phases during the cell cycle.

Mice lacking all interphase Cdks – Cdk2/Cdk4/Cdk6 *null* mice on a genetic background deficient for Cdk3 – died during midgestation, due to the same defect observed in animals lacking Cdk4 and Cdk6. The absence of Cdk2 mildly exacerbated this defect, since the triple knock-out animals died a couple of days (E12.5–E12.5) before those only lacking Cdk4 and Cdk6. However, deletion of Cdk2 in the absence of Cdk4 and Cdk6 did not result in additional defects, with the exception of a decrease in the number of heart cardiomyocytes, a defect also observed in newborn mice lacking Cdk2 and Cdk4 (Berthet et al. 2006; Barrière et al. 2007). These observations indicate that the interphase Cdks do not have compensatory activities in most cell types.

Finally, it is important to underscore that the full complement of interphase Cdks cannot compensate for Cdk1 activity, since embryos lacking this catalytic subunit can only accomplish a single round of cell division (Table 1; Santamaría et al. 2007 and our unpublished observations). Moreover, replacement of Cdk1 by Cdk2 by homologous recombination, even in the presence of the full complement of inter-phase Cdks, also resulted in early embryonic lethality (Satyanarayana et al. 2008), indicating that Cdk1 has unique properties that cannot be carried out by interphase Cdks even when expressed from the *Cdk1* locus.

4 Interphase Cdks Have Evolved to Drive Proliferation of Highly Specialized Cells

Why have multicellular eukaryotes evolved an increasing number of interphase Cdks? If they are not essential for successfully driving the basic cell cycle, they must have other essential roles that justify their appearance during evolution from unicellular to multicellular organisms. As indicated above, genetic analysis of Cdk function in the mouse has illustrated the need for specific Cdks to sustain adequate proliferation levels in specialized cells (Table 1). For instance, adult, but not embryonic, pancreatic beta cells have an absolute requirement for Cdk4 (Martín et al. 2003; Mettus and Rane 2003). Ectopic expression of Cdk6 in these cells restores proliferation (S. Ortega, personal communication). Yet, neither Cdk2 nor Cdk1 can compensate for the absence of Cdk4 in these cells. The most plausible explanation for these observations is that certain specialized cells control their homeostatic numbers by expressing specific Cdk substrates that control commit-ment for cell division (Fig. 2).

Likewise, Cdk4 is required to drive proliferation of erythroid precursors and embryonic cardiomyocytes. However, in these specialized cell types Cdk4 activity is compensated by Cdk6 and Cdk2, respectively (Malumbres et al. 2004; Berthet et al. 2006; Barrière et al. 2007). These compensatory specificities cannot be explained by the specific roles attributed to these kinases by the classical model of the cell cycle. Indeed, even quantitative models that postulate that the various interphase Cdks are required to reach a certain threshold of kinase activity are difficult to invoke to explain these observations. Hence, the most likely explanation is that specialized cell types have evolved additional controls to regulate their commitment to cell division. At least some of these controls may consist of molecular gatekeepers that require phosphorylation by selective interphase to allow these specialized cells to proceed through the cell cycle. Yet, such a hypoth-esis needs to be validated by experimental evidence.

Cdk2 poses a special paradox to explain its putative role in the mitotic cell cycle. This kinase binds efficiently to the E-type Cyclins, the most tightly regulated Cyclins, at least during interphase. Moreover, Cdk2 has been involved in countless phosphorylation processes, from inactivation of pocket proteins to centrosome duplication (Malumbres and Barbacid 2005). Yet, mice lacking Cdk2 are perfectly normal, expect in their germ cells (Berthet et al. 2003; Ortega et al. 2003). Is it possible that Cdk2 indeed participates in DNA synthesis – although genetics has clearly ruled out that it is essential – but cells have created a massive redundancy system by which any other Cdk, including Cdk1, can compensate for the absence of Cdk2. If so, why has such a redundancy system not been created for meiosis, a process equally critical for evolution and in which Cdk2 is absolutely essential? The most likely explanation is that Cdk2 was generated during evolution to perform an essential role in meiosis. Yet, organisms did not develop transcriptional regulatory mechanisms to limit Cdk2 expression to germ cells and allow Cdk2 to be expressed in mitotic cells. But why? We can only speculate that either Cdk2 is not detrimental to mitotic cells or it perhaps contributes in some non-essential way to the mitotic cell cycle that cannot be easily identified by genetic approaches.

5 Cell Cycle Cdks and Cancer

The mutations in Cdks and their regulators that contribute to human cancer have been well defined. Deregulation of CDK4 and CDK6 activities, but not of CDK2, has been implicated in a wide variety of human tumors (Ortega et al. 2002). CDK4 is altered in a small set of melanoma patients by a miscoding mutation (Arg24 to Cys) that blocks binding of INK4 inhibitors (Wölfel et al. 1995). CDK6 is over-expressed in some leukemias as a consequence of nearby translocations. *CDK4* and *CDK6* are amplified or overexpressed in several malignancies (sarcoma, glioma, breast tumors, lymphoma, melanoma, etc.), although the causal role of these alterations in tumor development is difficult to assess since at least *CDK4* is co-amplified with *MDM2* in most of these tumors (Malumbres and Barbacid 2001). On the other hand, misregulation of D-type Cyclins and INK4 inhibitors is a common feature of most tumor types, suggesting that the CDK4/6 kinases are hyperactive in most types of human cancers, with certain preference for CDK6 in mesenchymal tumors (leukemias and sarcomas) and CDK4 in epithelial malignancies (endocrine tissues and mucosas) as well as in some sarcomas. Experimental deregulation of the cell cycle in mice also elicits tumor development. For instance, constitutive activation of Cdk4 in a knock-in strain that carries the miscoding mutation (Arg24 to Cys) found in melanoma patients results in endocrine neoplasias (insulinomas, Leydig cell and pituitary tumors), epithelial (liver, gut, breast) hyperplasias and sarcomas, albeit after long latencies (Sotillo et al. 2001a; Rane et al. 2002). Interestingly, these mice do not develop melanoma unless insulted with a skin carcinogen (Sotillo et al. 2001b).

So far there are no models for Cdk6-induced tumorigenesis. CDK2 has not been found mutated in human cancer and there are no models of Cdk2-mediated tumors in experimental animals. Yet, expression of the p21^{CIP1} and p27^{KIP1} inhibitors is often lost during human tumor development and ablation of these inhibitors in the germ line of mice also results in tumor development (Malumbres and Barbacid 2001). Interestingly, genetic experiments have indicated that Cdk2 does not play a significant role in tumors lacking these inhibitors, in spite of been considered its primary target. Indeed, Martín et al. 2005 illustrated that ablation of Cdk2 had no effect on either latency or incidence of pituitary tumors generated by loss of p27^{Kip1}. Since p21^{Cip1} and p27^{Kip1} are also good inhibitors of Cdk1, it is possible that deregulation of Cdk1 activity might be responsible for tumor development in those malignancies lacking p21^{Cip1} or p27^{Kip1} expression. No mutations affecting Cdk1 have been identified, probably due to the central role that this kinase plays in the cell cycle.

6   Are Interphase Cdks Required for Tumor Development?

So far, we do not know to what extent interphase Cdks are essential to drive proliferation of tumor cells, except in those tumors in which they are mutated or clearly deregulated. The specific requirements of interphase Cdks for proliferation of specific cell lineages in mice raise the possibility that these Cdks may also be required for proliferation of tumor cells derived from those lineages. For instance, could we target insulinomas by inhibiting Cdk4, based on the fact that this kinase is essential for normal beta cell proliferation? This is an obvious question that, as yet, has not been addressed.

More interesting is the possibility that interphase Cdks may be required to sustain tumor growth derived from certain cell types even though they may not be required to maintain their normal homeostasis. This relevant question is beginning to be addressed by selectively ablating interphase Cdks in defined mouse tumor models. For instance, Cdk4 *null* mice, unlike their wild type counterparts, did not develop mammary tumors driven by overexpression of the HER2 receptor (Reddy et al. 2005; Yu et al. 2006). Interestingly, expression of Cdk4 did not appear to be necessary for mammary gland development. These results suggest that CDK4 inhibition by small molecules may have significant therapeutic efficacy in treating HER2-positive breast tumors, with limited toxicity. We have observed similar results in non-small cell lung carcinoma induced by an endogenous K-Ras^{G12V} oncogene. Induction of K-Ras^{G12V} expression in Cdk4-deficient mice, but not Cdk6 or Cdk2 mutant animals, resulted in a significant decrease in tumor development compared to wild type mice. Moreover, none of the observed tumors were of high grade (Puyol et al. 2010). Use of a conditional allele for *Cdk4* has allowed us to ask whether Cdk4 is required for tumor progression in pre-existing tumors. Tumor-bearing mice, as determined by computed tomography (CT) scanning, were inoculated with adenoviruses expressing Flpe recombinase (Ad-*Flpe*) to eliminate the conditional *Cdk4*^{frt} alleles. Whereas tumors of control mice developed rapidly, Ad-Flpe-treated tumors did not progress and only a few additional tumors were observed by CT scans (Fig. 4). More importantly, all tumors that progressed after Ad-*Flpe* exposure retained Cdk4 expression, due to lack of excision of the conditional allele (Fig. 4; Puyol et al. 2010).

A selective Cdk4/6 inhibitor, PD0332991, has been shown to have inhibitory activity against some human tumor xenografts that retained pRb expression (Fry et al. 2004). K-Ras^{G12V}-expressing mice carrying lung tumors, as determined by CT imaging, were treated either with vehicle or with two different doses of

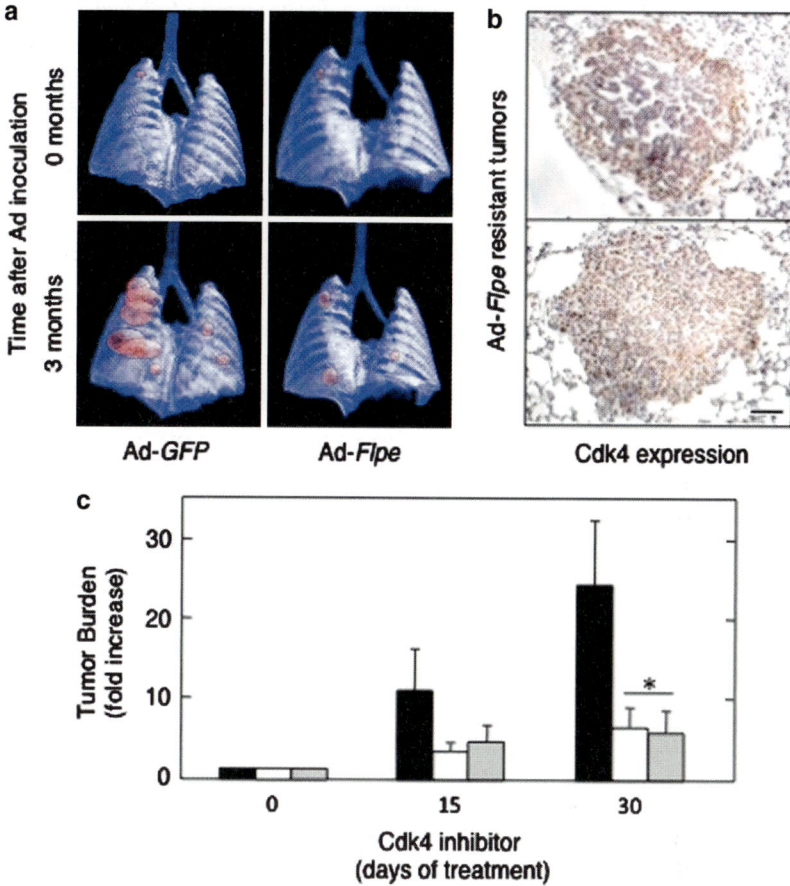

Fig. 4 Genetic and pharmacologic inhibition of expression block progression of K-*Ras* oncogene-driven NSCLCs. (**a**) Three-dimensional reconstruction of CT scans from representative K-*Ras*^{G12V}-expressing mice treated with adenoviral particles expressing control GFP (Ad-*GFP*) or Flpe recombinase (Ad-*Flpe*) to eliminate the conditional *Cdk4*^{frt} alleles. (**b**) Partial (*top*) and uniform (*bottom*) Cdk4 expression, as determined by IHC analysis with anti-Cdk4 polyclonal antibodies, in residual tumors of K-*Ras*^{G12V}-expressing mice treated with Ad-*Flpe*. Scale bar: 100 μm. (**c**) Increased tumor burden, determined by CT analysis, during a 30-day period in five-month-old tumor-bearing mice either left untreated (*solid bars*) or treated with an oral dose of 100 mg/kg (open bars) or 150 mg/kg (*gray bars*) of the PD0332991 Cdk4/6 selective inhibitor. Data shown as mean ± s.d. *P = 0.045

PD0332991 (the MTD of 150 mg/kg and a lower dose of 100 mg/kg) for 30 days. Mice were monitored for tumor development by CT to determine tumor burden 15 and 39 days after initiation of treatment. As shown in Fig. 4c, mice exposed to vehicle increased their tumor burden by 10-fold after 15 days of observation and 25-fold by the end of the experiment. In contrast, mice exposed to PD0332991 only increased their tumor burden by three- to four-fold at 15 days and five- to six-fold at 30 days, regardless of the dose. These results indicate that Cdk4 expression is an essential requirement for progression of already established K-Ras^{G12V}-driven tumors and suggest that the use of selective Cdk4 inhibitors may have therapeutic benefit for non small cell lung carcinoma (NSCLC) patients (Puyol et al. 2010).

Campaner et al. (2010) have recently illustrated that Cdk2 inhibits cellular senescence induced by the Myc oncogene. These observations were extrapolated to in vivo tumor systems using Eμ-*Myc* transgenic mice, a strain that express Myc in the B⁻cell lineage. In young Eμ-*Myc* mice at the pre-tumoral stage, Myc-induced changes in B⁻cell populations were comparable in normal and *Cdk2 null* animals. However, about one third of Eμ-*Myc* animals lacking Cdk2 showed elevated senescence, which correlated with lower levels of proliferation. Long-term follow-up of these mice revealed a statistically significant delay in the onset of lymphoma in the Eμ-*myc* animals lacking Cdk2, relative to their Cdk2-expressing siblings (Campaner et al. 2010). The relevance of this mechanism may depend on the tumor type, since Cdk2 deficiency did not delay Myc-induced oral carcinomas. Use of selective Cdk2 inhibitors also induced Myc-dependent senescence in rat embryo fibroblasts as well as in a human tumor cell line. Whether Cdk inhibitors will effectively inhibit Myc-driven tumors in vivo remains to be determined.

Why have CDK inhibitors displayed such modest activity in the clinic? It is possible that available compounds may have off-target effects that prevent them from reaching therapeutic concentrations. Whereas expected toxicities for CDK4 (diabetes), CDK2 (sterility) or CDK6 (mild anemia) inhibitors may be acceptable for adult cancer patients, inhibition of CDK1 (and possibly against CDK7) may hamper overall cell proliferation, resulting in toxicities not too different from those induced by current cytotoxics. It is therefore not surprising that promiscuous CDK inhibitors often display high toxicity in early clinical trials (Shapiro 2006). In addition, these inhibitors may produce other side effects by inhibiting less-studied kinases. To date, most CDK inhibitors undergoing clinical trials display limited specificity, with the exception of some compounds that selectively target CDK4 and CDK6 (Malumbres et al. 2008). As discussed above, it is possible that selective CDK4 inhibitors may have some therapeutic activity against HER2-positive breast tumors or K-Ras driven NSCLCs. Unfortunately, Cdk4/6 inhibitors have so far only been tested against lymphomas (Malumbres et al. 2008). Systematic inhibition of CDK activity by genetic means in relevant mouse tumor models may unveil unexpected synthetic lethal interactions between specific oncogenic signals and loss of selective interphase Cdks, such as those described above. These findings should provide valuable information to validate the potential uses of CDK inhibitors in cancer therapy.

Acknowledgments This work was funded by grants from the Spanish Ministry of Science and Innovation (MICINN) (SAF2006-11773 and Consolider-Ingenio 2010, CSD2007-00017), the 7th Framework Programme of the European Union (CHEMORES LSHG-CT-2007-037665) and the European Research Council (Advanced Grant ERC-2009-AdG_20090506).

References

Aleem E, Kiyokawa H, Kaldis P (2005) Cdc2-cyclin E complexes regulate the G1/S phase transition. Nat Cell Biol 7:831–836

Atanasoski S, Boentert M, De Ventura L, Pohl H, Baranek C, Beier K, Young P, Barbacid M, Suter U (2008) Postnatal Schwann cell proliferation but not myelination is strictly and uniquely dependent on cyclin-dependent kinase 4 (cdk4). Mol Cell Neurosci 37:519–527

Barrière C, Santamaría D, Cerqueira A, Galán J, Martín A, Ortega S, Malumbres M, Dubus P, Barbacid M (2007) Mice thrive without Cdk4 and Cdk2. Mol Oncol 1:72–83

Berthet C, Aleem E, Coppola V, Tessarollo L, Kaldis P (2003) Cdk2 knockout mice are viable. Curr Biol 13:1775–1785

Berthet C, Klarmann KD, Hilton MB, Suh HC, Keller JR, Kiyokawa H, Kaldis P (2006) Combined loss of Cdk2 and Cdk4 results in embryonic lethality and Rb hypophosphorylation. Dev Cell 10:563–573

Campaner S, Doni M, Hydbring P, Verrecchia A, Bianchi L, Sardella D, Schleker T, Perna D, Tronnersjö S, Murga M, Fernandez-Capetillo O, Barbacid M, Larsson LG, Amati B (2010) Cdk2 suppresses cellular senescence induced by the c-myc oncogene. Nat Cell Biol 12:54–59

Classon M, Harlow E (2002) The retinoblastoma tumour suppressor in development and cancer. Nat Rev Cancer 2:910–917

Drosten M, Dhawahir A, Sum EY, Urosevic J, Lechuga CG, Esteban LM, Castellano E, Guerra C, Santos E, Barbacid M (2010) Genetic analysis of Ras signalling pathways in cell proliferation, migration and survival. EMBO J 29:1091–104

Fry DW, Harvey PJ, Keller PR, Elliott WL, Meade M, Trachet E, Albassam M, Zheng X, Leopold WR, Pryer NK, Toogood PL (2004) Specific inhibition of cyclin-dependent kinase 4/6 by PD 0332991 and associated antitumor activity in human tumor xenografts. Mol Cancer Ther 3:1427–1438

Kastan MB, Bartek J (2004) Cell-cycle checkpoints and cancer. Nature 432:316–323

Kops GJ, Weaver BA, Cleveland DW (2005) On the road to cancer: aneuploidy and the mitotic checkpoint. Nat Rev Cancer 5:773–785

Kozar K, Ciemerych MA, Rebel VI, Shigematsu H, Zagozdzon A, Sicinska E, Geng Y, Yu Q, Bhattacharya S, Bronson RT, Akashi K, Sicinski P (2004) Mouse development and cell proliferation in the absence of D-cyclins. Cell 118:477–491

Malumbres M, Barbacid M (2001) To cycle or not to cycle: a critical decision in cancer. Nat Rev Cancer 1:222–231

Malumbres M, Barbacid M (2005) Mammalian cyclin-dependent kinases. Trends Biochem Sci 30:630–641

Malumbres M, Sotillo R, Santamaría D, Galán J, Cerezo A, Ortega S, Dubus P, Barbacid M (2004) Mammalian cells cycle without the D-type cyclin-dependent kinases Cdk4 and Cdk6. Cell 118:493–504

Malumbres M, Pevarello P, Barbacid M, Bischoff JR (2008) CDK inhibitors in cancer therapy: what is next? Trends Pharmacol Sci 29:16–21

Martín J, Hunt SL, Dubus P, Sotillo R, Néhmé-Pélluard F, Magnuson MA, Parlow AF, Malumbres M, Ortega S, Barbacid M (2003) Genetic rescue of Cdk4 null mice restores pancreatic beta-cell proliferation but not homeostatic cell number. Oncogene 22:5261–5269

Martín A, Odajima J, Hunt SL, Dubus P, Ortega S, Malumbres M, Barbacid M (2005) Cdk2 is dispensable for cell cycle inhibition and tumor suppression mediated by p27(Kip1) and p21 (Cip1). Cancer Cell 7:591–598

Massague J (2004) G1 cell-cycle control and cancer. Nature 432:298–306

Mettus RV, Rane SG (2003) Characterization of the abnormal pancreatic development, reduced growth and infertility in Cdk4 mutant mice. Oncogene 22:8413–8421

Moons DS, Jirawatnotai S, Parlow AF, Gibori G, Kineman RD, Kiyokawa H (2002) Pituitary hypoplasia and lactotroph dysfunction in mice deficient for cyclin-dependent kinase-4. Endocrinology 143:3001–3008

M. Barbacid

Nurse P (1997) The Josef Steiner Lecture: CDKs and cell-cycle control in fission yeast: relevance
    to other eukaryotes and cancer. Int J Cancer 71:707–708
Ortega S, Malumbres M, Barbacid M (2002) Cyclin D-dependent kinases, INK4 inhibitors and
    cancer. Biochim Biophys Acta 1602:73–87
Ortega S, Prieto I, Odajima J, Martín A, Dubus P, Sotillo R, Barbero JL, Malumbres M,
    Barbacid M (2003) Cyclin-dependent kinase 2 is essential for meiosis but not for mitotic cell
    division in mice. Nat Genet 35:25–31
Puyol M, Martín A, Dubus P, Mulero F, Pizcueta P, Khan G, Guerra C, Santamaría S, Barbacid M
    (2010) A synthetic lethal interaction between K-Ras oncogenes and Cdk4 unveils a therapeutic
    strategic for non small cell lung carcinoma. Cancer Cell 18:63–73
Rane SG, Dubus P, Mettus RV, Galbreath EJ, Boden G, Reddy EP, Barbacid M (1999) Loss of
    Cdk4 expression causes insulin-deficient diabetes and Cdk4 activation results in beta-islet cell
    hyperplasia. Nat Genet 22:44–52
Rane SG, Cosenza SC, Mettus RV, Reddy EP (2002) Germ line transmission of the Cdk4(R24C)
    mutation facilitates tumorigenesis and escape from cellular senescence. Mol Cell Biol
    22:644–656
Reddy HK, Mettus RV, Rane SG, Graña X, Litvin J, Reddy EP (2005) Cyclin-dependent kinase
    4 expression is essential for neu-induced breast tumorigenesis. Cancer Res 65:10174–10178
Santamaría D, Barrière C, Cerqueira A, Hunt S, Tardy C, Newton K, Cáceres JF, Dubus P,
    Malumbres M, Barbacid M (2007) Cdk1 is sufficient to drive the mammalian cell cycle.
    Nature 448:811–815
Satyanarayana A, Berthet C, Lopez-Molina J, Coppola V, Tessarollo L, Kaldis P (2008) Genetic
    substitution of Cdk1 by Cdk2 leads to embryonic lethality and loss of meiotic function of Cdk2.
    Development 135:3389–3400
Shapiro GI (2006) Cyclin-dependent kinase pathways as targets for cancer treatment. J Clin Oncol
    24:1770–1783
Sotillo R, Dubus P, Martín J, de la Cueva E, Ortega S, Malumbres M, Barbacid M (2001a) Wide
    spectrum of tumors in knock-in mice carrying a Cdk4 protein insensitive to INK4 inhibitors.
    EMBO J 20:6637–6647
Sotillo R, García JF, Ortega S, Martin J, Dubus P, Barbacid M, Malumbres M (2001b) Invasive
    melanoma in Cdk4-targeted mice. Proc Natl Acad Sci USA 98:13312–13317
Tsutsui T, Hesabi B, Moons DS, Pandolfi PP, Hansel KS, Koff A, Kiyokawa H (1999) Targeted
    disruption of CDK4 delays cell cycle entry with enhanced p27(Kip1) activity. Mol Cell Biol
    19:7011–7019
Wölfel T, Hauer M, Schneider J, Serrano M, Wölfel C, Klehmann-Hieb E, De Plaen E, Hankeln T,
    Meyer zum Büschenfelde KH, Beach D (1995) A p16INK4a-insensitive CDK4 mutant tar-
    geted by cytolytic T lymphocytes in a human melanoma. Science 269:1281–1284
Ye X, Zhu C, Harper JW (2001) A premature-termination mutation in the Mus musculus cyclin-
    dependent kinase 3 gene. Proc Natl Acad Sci USA 98:1682–1686
Yu Q, Sicinska E, Geng Y, Ahnström M, Zagozdzon A, Kong Y, Gardner H, Kiyokawa H, Harris
    LN, Stål O, Sicinski P (2006) Requirement for CDK4 kinase function in breast cancer. Cancer
    Cell 9:23–32

The Role of Cdk5 as a Cell Cycle Suppressor in Post-mitotic Neurons

Karl Herrup

Abstract Neurons of the central nervous system (CNS) leave the mitotic cycle when they leave the ventricular zone during embryonic and early postnatal development. Normally, they will never re-enter the cell cycle for the rest of the life of the organism. This rule is now known to be broken in many types of neurodegenerative disease. In these situations, nerve cells at risk for death have greatly elevated expression of cell cycle-related proteins; they have also been found to replicate their DNA. The existence of this pathway to neuronal death through the cell cycle raises the question of how a normal adult neuron suppresses cell division and places high therapeutic value on encouraging the activity of those proteins involved in the process. We have developed several lines of evidence that cyclin-dependent kinase 5 (Cdk5) is one such protein. To function as a cell cycle suppressor, Cdk5 must be located in the nucleus and it must be able to bind its cyclin-like activator, p35. Curiously, however, it does not need to retain kinase activity. Instead, its activity derives from its ability to sequester the E2F1 transcription factor and block its access to the DP1 co-factor, which greatly reduces binding to various cell cycle protein gene promoters thus inhibiting the cycle. Cdk5 stands as an excellent example of proteins whose functions are needed for the regulation of both differentiation and cell division. From this description of dual-specificity proteins, a concept is presented that the processes of division and differentiation are not so much independent as overlapping analog functions that must be balanced both during development and in the adult. The loss of balance would be expected to lead to neurodegeneration in a neuron or cancer in a less highly differentiated cell type.

K. Herrup
Department of Cell Biology and Neuroscience, Rutgers University, 604 Allison Road, Piscataway, NJ 08854, USA
e-mail: herrup@biology.rutgers.edu

T. Curran and Y. Christen (eds.), *Two Faces of Evil: Cancer and Neurodegeneration*, Research and Perspectives in Alzheimer's Disease, DOI 10.1007/978-3-642-16602-0_2, © Springer-Verlag Berlin Heidelberg 2011

1   Introduction

The control of the cell cycle control is a critical capability for all forms of life. From the simplest unicellular organism to the most complex mammal, survival depends on the ability to divide when possible and to stop dividing when conditions require. For a single cell organism such as a yeast cell or a bacterium, the regulation seems fairly straightforward: divide when nutrients are present and stop when they are not. For a neuron in the CNS of an adult human being, the regulatory constraints are more complex. During the early neurogenic phases of development, the progenitors of the various neuronal cell types must divide quickly and consistently. During the final stages of neurogenesis, a pattern of asymmetric divisions is established to allow the generation of the specialized cells that will populate the various brain regions in the adult. As most neurons in the adult CNS complete their last cell division during this period of embryogenesis and early postnatal life, a final challenge faced by the typical adult neuron is how to hold the cell cycle in check for the rest of the life of the organism. The need to do so is clear, as many laboratories have shown that, in different neurodegenerative diseases, the appearance of cell cycle events in mature neurons – re-expression of cell cycle proteins and the replication of DNA – are a near certain harbinger of impending cell death. Because of this tight linkage to disease, we have been engaged in a search for proteins whose functions include cell cycle suppression in mature neurons. Surprisingly, each of these proteins has also been identified as performing a highly specific function in the physiological health of a normal neuron. The enzyme known as cyclin dependent kinase (Cdk5) is one such protein.

2   Cdk5 Plays an Important Role in Neuronal Development

Cdk5 was identified as a member of the Cdk5 family of cell cycle-related kinases by virtue of its ability to bind typical cyclins such as cyclin D (Xiong et al. 1992). It was soon assigned to the status of 'atypical' Cdk, as overexpression in cultured cells did not drive the cell cycle. Further, the binding of cyclins such as cyclin D did not enhance kinase activity. The gene for the kinase was subsequently shown to be expressed in the maturing regions of the embryonic nervous system, not in the ventricular zone (Tsai et al. 1993), which further re-enforced the idea that Cdk5 was not a true cell cycle kinase. Rather, reduced Cdk5 activity was found to impede neuronal process growth (Nikolic et al. 1996). Subsequent studies of an engineered null mutation of Cdk5 showed that embryos lacking Cdk5 activity die at the end of embryogenesis, with massive failures of neuronal cell migration and cytological maturation (Gilmore et al. 1998; Ohshima et al. 1996). Cdk5 has two potential cyclin-like activators, p35 and p39. While the genetic inactivation of one of these has cortical lamination defects (Chae et al. 1997), elimination of both genes produces a spectrum of developmental defects that is indistinguishable from that seen in the Cdk5 knockout embryos (Ko et al. 2001). This finding was yet another piece of evidence identifying Cdk5 as a kinase involved in differentiation and maturation rather than in cell cycle regulation. This differentiation activity of Cdk5, in particular its role in migration and neuronal development, has been largely ascribed to its ability to phosphorylate a variety of cytoskeletal proteins ranging from tau to neurofilament (Hosoi et al. 1995; Veeranna et al. 1995).

3 Cdk5 Serves Important Functions in the Regulation of Synaptic Function

As the roles of Cdk5 in post-mitotic neuronal activity were explored further, several labs reported evidence for its involvement in the phosphorylation of numerous synaptic proteins (Bibb 2003; Cheng and Ip 2003; Fischer et al. 2003; Smith and Tsai 2002; Tan et al. 2003). The list of synaptic proteins whose activity is modified by Cdk5 phosphorylation has grown year by year. The synaptic vesicle phospho-protein, synapsin I, was identified early on (Matsubara et al. 1996). Later, proteins such as Munc-18, dynamin, amphphysin and many others were also shown to be synaptic targets. Functional changes in synaptic functions are clearly seen when these phosphorylations are prevented or enhanced. As only one example, Cdk5 phosphorylates TrkB, and blocking Cdk5 activity essentially blocks brain-derived neurotrophic factor (BDNF)-triggered dendritic growth in primary hippocampal neurons (Cheung et al. 2007).

4 Cdk5 is a Cell Cycle Suppressor in Post-mitotic Neurons

Inspired by the appearance of Cdk5 and its cyclin-like activator, p35, immediately after the neuroblasts of the CNS leave the ventricular zone, my laboratory has explored the possible role that Cdk5 might play a role in inhibiting rather than advancing the cell cycle. The first suggestion that this might be true came from a re-examination of the phenotype of $Cdk5^{-/-}$ embryonic cortical neurons both in vivo and in vitro (Cicero and Herrup 2005). The migration of most Cdk5-deficient neurons is abnormal, but even those neurons that do achieve normal or near normal positions are defective in their development. This is made most apparent by examining their biochemical maturation. Nestin, a commonly used marker of mitotic neuronal progenitor cells, is robustly expressed in $Cdk5^{-/-}$ neurons, even in those that have migrated substantial distances from the ventricular zone. In keeping with this biochemical evidence of an immature state of existence, the levels of Map2 staining, a commonly used marker of mature or maturing dendrites, is virtually absent. This same block in maturation can be found when neurons from mutant and wild type embryos are cultured in vitro. Wild type embryos lose the immature marker, TuJ1, and increase expression of Map2; $Cdk5^{-/-}$ neurons never lose their TuJ1-positive status and fail to show any increase in Map2 staining (Cicero and Herrup 2005). This biochemical evidence for a blocked maturation and persistent precursor status was also seen in the ability of $Cdk5^{-/-}$ neurons to regulate their cell cycle. Both in vivo and in vitro, wild type neurons show little or no evidence of cell cycle involvement, whereas $Cdk5^{-/-}$ neurons cycle and these cycling neurons can be shown to die (Cicero and Herrup 2005).

5   Cdk5 Inhibits the Cell Cycle in a Kinase-Independent Fashion

The loss of cell cycle control in the Cdk5-deficient neurons suggests that phosphorylation of one or more Cdk5 target substrates is normally required to hold the neuronal cell cycle in check. A potential candidate for just such a substrate was the tumor suppressor gene, retinoblastoma (RB), a recognized target of Cdk5 (Hamdane et al. 2005; Lee et al. 1997). This hypothesis, however, proved incorrect. Culturing of wild type neurons in the presence of roscovitine, a Cdk5 inihibitor, reduced their dendritic complexity but did not release the cell cycle (Cicero and Herrup 2005). Further transfection of a kinase-dead (KD) form of Cdk5 into $Cdk5^{-/-}$ neurons was completely effective in restoring full cell cycle control to the mutant neurons (Zhang et al. 2008).

This unexpected set of observations prompted us to examine the behavior of Cdk5 during a typical cell cycle. We verified earlier findings that total Cdk5 levels do not change with cell cycle arrest (by serum withdrawal) or release (by restoration of serum). The subcellular localization of Cdk5, however, changes dramatically. In the arrested cells, Cdk5 is predominantly nuclear; in the cycling cells, Cdk5 is predominantly cytoplasmic (Zhang et al. 2008), suggesting that it was Cdk5 in the nucleus that was the key to neuronal cell cycle suppression. We validated this finding by showing that both wild type Cdk5 and Cdk5 with a nuclear localization signal (NLS) were capable of arresting the incorporation of BrdU in $Cdk5^{-/-}$ neurons. By contrast, Cdk5 with a nuclear export signal (NES) was unable to rescue the mutant cell cycle. To prove that nuclear export was the instigating factor, we stimulated wild type neurons to cycle and die with fibrilarized $A\beta_{1-42}$ peptide or $A\beta$-stimulated microglial conditioned medium (Wu et al. 2000). In the presence of the nuclear export inhibitor, leptomycin-B, the $A\beta$ peptide could not induce a neuronal cell cycle (Zhang et al. 2008).

These experiments identified the importance of nuclear localization for Cdk5 cell cycle suppression but did not speak directly to the requirements for kinase activity. To approach this problem, we turned first to the neuroblastoma cell line, N2a. When wild type Cdk5 is transfected into log phase N2a cells, the level of BrdU incorporation decreases by five to ten-fold. Nuclear Cdk5 (Cdk5–NLS) is equally effective, but Cdk-NES was inactive. With Cdk5-NES, cell cycle activity was as high as with control GFP-only transfections (Zhang et al. 2010). We repeated these experiments with a KD form of Cdk5 (Nikolic et al. 1996; Tsai et al. 1994) and validated our earlier findings. KD Cdk5 was as effective as wild type at inhibiting the N2a cell cycle. As with the wild type protein, KD–Cdk5–NLS (nuclear) was a cell cycle suppressor whereas KD–Cdk5–NES (cytoplasmic) was not. This effect was not restricted to neuroblastoma cells. The same set of findings was seen in the ventricular zone of the E14.5 mouse embryo after in utero electroporation of GFP–Cdk5 or its KD isoform (Zhang et al. 2010).

6 Cdk5 Inhibits the Cell Cycle by Sequestering E2F1

These findings leave unanswered the mechanistic question of how nuclear Cdk5 is able to so effectively arrest the cell cycle. We turned to the E2F1 protein for answers. We first used an electrophoretic mobility shift assay (EMSA) to show that, in the presence of Cdk5, occupancy of E2F1 response elements by E2F1 was dramatically reduced (Zhang et al. 2010). We subsequently demonstrated that E2F1 is physically associated with Cdk5 both in situations where the two proteins are overexpressed – in cell lines or neurons – and when co-immunoprecipitations are performed from whole brain.

The association of Cdk5 and E2F1 led us to ask whether there might be any other proteins associated with the complex. We had determined that Cdk5 did not require its kinase activity to block the cycle (Cicero and Herrup 2005; Zhang et al. 2008), but we wished to determine whether Cdk5 required the presence of p35 to form a complex with E2F1. We engineered a p35 binding-deficient Cdk5 isoform (Tarricone et al. 2001) and repeated our N2a and in utero electroporation experiments. To our surprise, the p35 binding-deficient Cdk5 (Δ35) was ineffective in blocking the cell cycle, suggesting that p35 was a part of the Cdk5-E2F1 complex. We verified this suggestion by co-immunoprecipitation of p35 and E2F1. The critical nature of the presence of p35 for the formation of the complex was shown by the observation that Cdk5-Δ35 is unable to associate with E2F1 when the two are overexpressed in the presence of p35.

Further insights into the dynamics of the complex were achieved when we examined the behavior of the E2F1 DNA binding co-factor, DP1. In the absence of DP1, the cell cycle stimulatory activity of E2F1 is decreased significantly (Bandara et al. 1993). We showed that the presence of Cdk5 and p35 in the nucleus successfully displaces DP1 from its binding to E2F1, effectively diminishing the efficiency of E2F1 as a promoter binding element. This finding helps to explain the EMSA findings as well as the variety of cell cycle effects reported above. Curiously, when the same combination of proteins was expressed in the cytoplasm, the situation reversed. Now DP1 and E2F1 remained tightly bound whereas Cdk5 was excluded. Clearly there are additional insights left to be had.

7 The Implications of the Central Role of Cdk5 in Neuronal Cell Cycle Suppression

The story of Cdk5 has several important lessons to teach us about the relationships between cancer and neurodegeneration – the subject of this IPSEN meeting. It teaches us that Cdk5 may well be an atypical Cdk, but not because it has *no* role in the cell cycle. It is atypical primarily because its role is to suppress rather than encourage the cell cycle. It is also atypical because it achieves this activity without the use of its kinase activity yet it requires association with its cyclin-like partner, p35. The implication of these characteristics is that the well-known association of the Cdk family of kinases with their activating subunits (the cyclins) and other inhibitory elements (the Cip/Kip and Ink families of Cdk inhibitors) may well have functions beyond changing the extent to which Cdk target substrates become phosphorylated. The associations themselves may prove a significant part of the story of any Cdk.

Finding a role for Cdk5 in cell cycle activity should also remind us of a deeper lesson: cell cycle and cell differentiation are highly inter-related activities and they are often regulated by the same protein. An early example of this was the discovery of a role for the RB tumor suppressor in neuronal differentiation as well as cell cycle control (Lee et al. 1994). Perhaps the most recent example is that of the protein that is mutated in ataxia-telangiectasia, ATM. Long recognized as a cell cycle checkpoint response protein during the DNA damage response, ATM also plays a major role in maintaining the complex economy of neuronal synaptic vesicle functioning (Li et al. 2009). Many other examples of cell cycle proteins functioning in the development and differentiation of neurons have been discovered (reviewed in Frank and Tsai 2009). Cdk5 fits well into this category of dual-specificity proteins.

The discovery of growing numbers of proteins in this class suggests that it may be time to loosen the normally binary view of cell cycle and cell differentiation. A neuron that has left the ventricular zone of the CNS is viewed as permanently post-mitotic. Before it left the ventricular zone, it was dividing; after it left, it was differentiating. A more appropriate description of this event might be to assign both division and differentiation an analog value; this idea is crudely diagrammed in Fig. 1. Beginning with fertilization, the primary activity of the cells of the zygote is division. The mass of the organism must increase. As development proceeds (upwards in the figure), cells move from the totipotency and rapid cell cycle kinetics of the embryonic stem cell through more restricted progenitor cell phases and finally to a stage where terminal differentiation occurs. In most cell types, as this process unfolds, the pressure to divide decreases, as is shown by the lightening and narrowing of the red wedge labeled "Division". At the same time, the cells are increasingly expressing their differentiation program, indicated by the darkening and the broadening of the green wedge labeled "Differentiation" and the nervous system-specific labels such as "synapse". If we add the dimension of time specifically to this progression, a single neuron can trace its heritage through this lineage, as shown in Fig. 2.

Fig. 1 The program of development can be seen as two analog processes – division and differentiation – that wax and wane inversely to each other. As development proceeds (from the *bottom* to the *top* of the figure), the two processes change in relative proportion to each other, but neither is ever completely eliminated. ES, embryonic stem

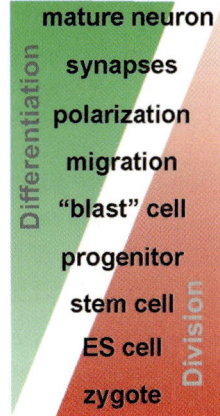

Fig. 2 The trajectory of the lineage of one adult neuron through the progression shown in Fig. 1. Developmental time is shown as the *rightward arrow*

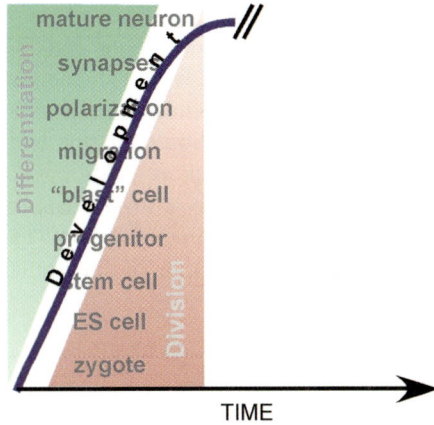

Both cancer and certain forms of neurodegenerative disease can be incorporated into this scheme (Fig. 3). In cancer, certain stem cells or blast-like cells lose their differentiation properties and slip back towards a state where cell division is more aggressive and hence more likely. The description of how this occurs through both genetic and epigenetic means will be left to others in this symposium. For neurodegeneration, the scheme reminds us that the populations of neurons that are at risk for death in diseases such as Alzheimer's, Parkinson's and others display substantial evidence of re-activating the machinery of the cell cycle. In the scheme shown in Fig. 3, this process can be envisioned as a mature neuron slipping downwards; the extent of differentiation decreases while the probability of division increases. At some tipping point, the neuron, while still highly differentiated, triggers a cell cycle, which is ultimately a lethal decision and the result is neurodegenerative disease.

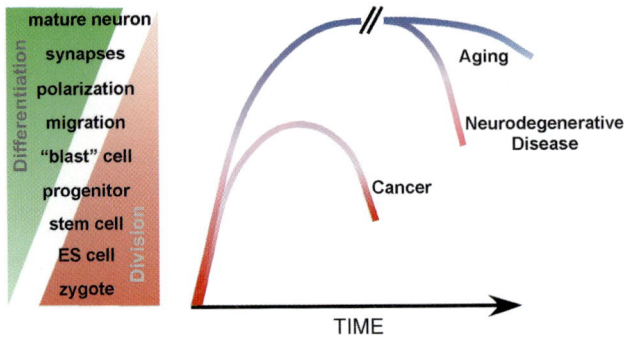

Fig. 3 On this continuum, neurodegenerative disease can be seen as a retreat from the highly differentiated state of the mature neuron, eventually leading to an enhanced susceptibility to cycle re-initiation and cell death. Cancer can be viewed as a similar failure, but usually from a less differentiated blast cell or stem-like state. The trajectory is similar, however: a descent into a state where susceptibility to the re-initiation of the cell cycle is enhanced

This concept of differentiation and division as two competing forces in development is not a new one. What the emerging evidence suggests, however, is that the competition is never ending. A differentiated cell, such as a neuron, must constantly and actively promote differentiation and suppress cell division. It may be speculated that proteins, such as Cdk5, have evolved to efficiently promote both processes. Taking a cue from these dual-specificity proteins, it may be that the most effective strategies for combating both cancer and neurodegeneration are those that involve both the promotion of differentiation and the inhibition of division.

References

Bandara LR, Buck VM, Zamanian M, Johnston LH, La Thangue NB (1993) Functional synergy between DP-1 and E2F-1 in the cell cycle-regulating transcription factor DRTF1/E2F. Embo J 12:4317–4324

Bibb JA (2003) Role of Cdk5 in neuronal signaling, plasticity, and drug abuse. Neurosignals 12: 191–199

Chae T, Kwon YT, Bronson R, Dikkes P, Li E, Tsai L-H (1997) Mice lacking p35, a neuronal specific activator of Cdk5, display cortical lamination defects, seizures, and adult lethality. Neuron 18:29–42

Cheng K, Ip NY (2003) Cdk5: a new player at synapses. Neurosignals 12:180–190

Cheung ZH, Chin WH, Chen Y, Ng YP, Ip NY (2007) Cdk5 is involved in BDNF-stimulated dendritic growth in hippocampal neurons. PLoS Biol 5:e63

Cicero S, Herrup K (2005) Cyclin-dependent kinase 5 is essential for neuronal cell cycle arrest and differentiation. J Neurosci 25:9658–9668

Fischer A, Sananbenesi F, Spiess J, Radulovic J (2003) Cdk5: a novel role in learning and memory. Neurosignals 12:200–208

Frank CL, Tsai LH (2009) Alternative functions of core cell cycle regulators in neuronal migration, neuronal maturation, and synaptic plasticity. Neuron 62:312–326

Gilmore EC, Ohshima T, Goffinet AM, Kulkarni AB, Herrup K (1998) Cyclin-dependent kinase 5-deficient mice demonstrate novel developmental arrest in cerebral cortex. J Neurosci 18: 6370–6377

Hamdane M, Bretteville A, Sambo AV, Schindowski K, Begard S, Delacourte A, Bertrand P, Buee L (2005) p25/Cdk5-mediated retinoblastoma phosphorylation is an early event in neuronal cell death. J Cell Sci 118:1291–1298

Hosoi T, Uchiyama M, Okumura E, Saito T, Ishiguro K, Uchida T, Okuyama A, Kishimoto T, Hisanaga S (1995) Evidence for cdk5 as a major activity phosphorylating tau protein in porcine brain extract. J Biochem (Tokyo) 117:741–749

Ko J, Humbert S, Bronson RT, Takahashi S, Kulkarni AB, Li E, Tsai LH (2001) p35 and p39 are essential for cyclin-dependent kinase 5 function during neurodevelopment. J Neurosci 21: 6758–6771

Lee EY, Hu N, Yuan SS, Cox LA, Bradley A, Lee WH, Herrup K (1994) Dual roles of the retinoblastoma protein in cell cycle regulation and neuron differentiation. Genes Dev 8: 2008–2021

Lee K, Helbing C, Choi K, Johnston R, Wang J (1997) Neuronal Cdc2-like kinase (Nclk) binds and phosphorylates the retinoblastoma protein. J Biol Chem 272:5622–5626

Li J, Han YR, Plummer MR, Herrup K (2009) Cytoplasmic ATM in neurons modulates synaptic function. Curr Biol 19:2091–2096

Matsubara M, Kusubata M, Ishiguro K, Uchida T, Titani K, Taniguchi H (1996) Site-specific phosphorylation of synapsin I by mitogen-activated protein kinase and Cdk5 and its effects on physiological functions. J Biol Chem 271:21108–21113

Nikolic M, Dudek H, Kwon Y, Ramos Y, Tsai L (1996) The cdk5/p35 kinase is essential for neurite outgrowth during neuronal differentiation. Genes Dev 10:816–825

Ohshima T, Ward JM, Huh CG, Longnecker G, Veeranna Pant HC, Brady RO, Martin LJ, Kulkarni AB (1996) Targeted disruption of the cyclin-dependent kinase 5 gene results in abnormal corticogenesis, neuronal pathology and perinatal death. Proc Natl Acad Sci USA 93:11173–11178

Smith DS, Tsai LH (2002) Cdk5 behind the wheel: a role in trafficking and transport? Trends Cell Biol 12:28–36

Tan TC, Valova VA, Malladi CS, Graham ME, Berven LA, Jupp OJ, Hansra G, McClure SJ, Sarcevic B, Boadle RA, Larsen MR, Cousin MA, Robinson PJ (2003) Cdk5 is essential for synaptic vesicle endocytosis. Nat Cell Biol 5:701–710

Tarricone C, Dhavan R, Peng J, Areces LB, Tsai LH, Musacchio A (2001) Structure and regulation of the CDK5-p25(nck5a) complex. Mol Cell 8:657–669

Tsai LH, Takahashi T, Caviness VS Jr, Harlow E (1993) Activity and expression pattern of cyclin-dependent kinase 5 in the embryonic mouse nervous system. Development 119:1029–1040

Tsai LH, Delalle I, Caviness VS Jr, Chae T, Harlow E (1994) p35 is a neuronal-specific regulatory subunit of cyclin-dependent kinase 5. Nature 271:419–423

Veeranna SKT, Link WT, Jaffe H, Wang J, Pant HC (1995) Neuronal cyclin-dependent kinase-5 phosphorylation sites in neurofilament protein (NF-H) are dephosphorylated by protein phosphatase 2A. J Neurochem 64:2681–2690

Wu Q, Combs C, Cannady SB, Geldmacher DS, Herrup K (2000) Beta-amyloid activated microglia induce cell cycling and cell death in cultured cortical neurons. Neurobiol Aging 21: 797–806

Xiong Y, Zhang H, Beach D (1992) D type cyclins associate with multiple protein kinases and the DNA replication and repair factor PCNA. Cell 71:505–514

Zhang J, Cicero SA, Wang L, Romito-Digiacomo RR, Yang Y, Herrup K (2008) Nuclear localization of Cdk5 is a key determinant in the postmitotic state of neurons. Proc Natl Acad Sci USA 105:8772–8777

Zhang J, Li H, Yabut OY, Fitzpatrick H, D'Arcangelo G, Herrup K (2010) Cdk5 suppresses the neuronal cell cycle by disrupting the E2F1-DP1 complex. J Neurosci 30:5219–5228

Actin-SRF Signaling in the Developing and Mature Murine Brain

Alfred Nordheim and Bernd Knöll

Abstract Functional characterization of the transcription factor Serum Response Factor (SRF) has generated new insight into cellular mechanisms by which the actin cytoskeletal microfilament communicates with the genome. In essence, the SRF cofactors of the Myocardin Related Transcription Factor (MRTF) protein family sense cytoplasmic G-actin levels such that, upon stimulated F-actin polymerization, MRTFs translocate to the nucleus. This translocation leads to activated transcription of SRF target genes, a subset of which encodes actin and other cytoskeletal proteins. Migratory activities of cells are modulated thereby, including motile functions of neurons and metastatic cancer cells. In the mouse brain, SRF-directed gene expression contributes to both embryonic and adult activities. Essential neuronal migratory events of pre- and postnatal brain development are impaired in SRF-deficient mouse mutants. Furthermore, in the mature brain, essential cognitive functions, including learning and memory, are critically dependent upon SRF and its MRTF cofactors.

1 Principles of Actin-SRF Signaling in Brain Cells

1.1 Actin Dynamics Regulates Gene Expression Upon Activation of MRTF–SRF Signaling

Dynamical rearrangements of the actin cytoskeleton drive a multitude of cellular behaviors, including migration, adhesion, and guided movement. These actin-driven cell activities require tightly coordinated de novo protein synthesis.

A. Nordheim (✉)
Vertebrate Gene Expression and Organ Function, Department of Molecular Biology, Interfaculty Institute for Cell Biology, University of Tübingen, Auf der Morgenstelle 15, 72076 Tübingen, Germany
e-mail: alfred.nordheim@uni-tuebingen.de

T. Curran and Y. Christen (eds.), *Two Faces of Evil: Cancer and Neurodegeneration*,
Research and Perspectives in Alzheimer's Disease, DOI 10.1007/978-3-642-16602-0_3,
© Springer-Verlag Berlin Heidelberg 2011

Underlying changes in gene expression profiles are, in part, controlled by the transcription factor SRF, which is regulated by functional interactions with cofactors (for review, see Olson and Nordheim 2010). Whereas the Ternary Complex Factor (TCF) family of SRF cofactors relays MAPK-mediated activation to a subset of SRF target genes (for review, see Buchwalter et al. 2004), members of the myocardin family of SRF cofactors communicate actin dynamics to SRF activity (for review, see Posern and Treisman 2006). The myocardin protein family includes myocardin and the myocardin-related transcription factors (MRTFs) A and B (also known as MAL/Mkl1 and Mkl2). In many cell types, MRTF proteins shuttle between the cytoplasm and the nucleus, dependent upon their release from a cytoplasmic G-actin/MRTF complex (Miralles et al. 2003; Vartiainen et al. 2007). SRF target genes controlled by MRTF shuttling encode, among others, cytoskeletal proteins (including actin itself), contractile proteins, and extracellular matrix-associated proteins. By knock-out mutagenesis in murine embryonic stem cells, mouse embryos, and newborn mice, we demonstrated the essential role of SRF in regulating cell migration, adhesion, neuronal growth, axonal outgrowth, and axon guidance (Schratt et al. 2002; Alberti et al. 2005; Knoll et al. 2006; for review, see Knoll and Nordheim 2009). Similarly, neuron-specific double knock-out of the *Mrtfa* and *Mrtfb* genes in the mouse genome causes impairments in both neuronal migration and neurite outgrowth (Mokalled et al. 2010). In vitro, MRTF-dependent neuronal motility requires SRF (Knoll et al. 2006). Conversely, SRF-stimulated gene expression is impaired by dominant-negative MRTF (Wickramasinghe et al. 2008). Thus, the actin-MRTF–SRF pathway emerges as an essential module to link cytoskeletal dynamics with nuclear gene activity in many types of cells, including neurons.

1.2 Actin Dynamics in Neuronal Network Assembly

During initial wiring of neuronal networks, neurons pass through stages of morphological differentiation, intimately linked to cellular motility and cytoskeletal dynamics (Barnes and Polleux 2009). Dynamic alterations of actin microfilaments together with other cytoskeletal structures (e.g., microtubules) provide an essential cellular framework to ensure such processes as neuronal cell migration, neurite outgrowth, axon vs. dendrite specification, axon guidance and synaptic targeting (Dent and Gertler 2003; Pak et al. 2008). In the adult nervous system, structural changes during generation of new synapses and their modulation in synaptic plasticity also rely on dynamic properties of the cytoskeleton.

In neurons, unlike many other cell types (e.g., fibroblasts), F-actin is not uniformly distributed in the entire cytoplasm. Instead, F-actin is primarily polymerized from G-actin subunits in the distant growth cones, the motile axon tips of neurons

(Dent and Gertler 2003; Pak et al. 2008). Within the growth cone, F-actin occupies peripheral growth cone areas and G-actin occupies central growth cone areas (Torreano et al. 2005). The growth cones containing most of the neuronal actin are often far away from the neuronal cell body, including the nucleus. Thus, unlike what is observed in non-neuronal cell types, where actin/MRTF is in close vicinity to the nucleus, in neurons, MRTF – as a functional 'sensor' of actin dynamics – might have to bridge a longer distance to communicate between the distal growth cone and the nucleus.

1.3   Actin-MRTF–SRF interplay in Neuronal Signaling

How does actin communicate with SRF to allow for coordinated neuronal activities? In principal, the general mechanism of an actin-MRTF–SRF interplay described for non-neuronal cells (see above) appears to also operate in neurons (Knoll and Nordheim 2009; Knoll 2010). We recently employed actin mutants (Stern et al. 2009) that were well characterized with regard to their polymerization properties (Posern et al. 2002, 2004) to explore the consequences of enhanced F- vs. G-actin levels on neuronal SRF-mediated gene expression and neuron morphology. Actin G15S, raising F-actin polymer abundance, enhanced SRF-dependent gene expression and neuronal motility (Stern et al. 2009). Mechanistically, increased F-actin density might, similar to fibroblasts, liberate MRTF for nuclear translocation followed by interaction with SRF. However in neurons (and also some tumor cell lines; Medjkane et al. 2009), cytoplasm-to-nucleus shuttling (Tabuchi et al. 2005; Wickramasinghe et al. 2008) has been reported as well as constitutive nuclear MRTF localization (Kalita et al. 2006; Stern et al. 2009). This discrepancy might be explained by differences in MRTF localization dependent upon cell type, anti-MRTF antisera used, and time-points investigated. Thus, from a spatial point of view, it will be interesting to explore how nucleus-resident MRTF is sensing actin alterations in the distant growth cone.

In contrast to G15S actin, R62D actin, a non-polymerizing mutant actin protein, blocked SRF-mediated gene expression and reduced neurite outgrowth (Stern et al. 2009). Of note, the cytoplasmic R62D actin mutant induces similar aberrant neuron morphology and reduces SRF gene expression when restricted exclusively to nuclear compartment, which was achieved by fusing R62D actin with a nuclear localization signal (NLS; Stern et al. 2009). This effect of R62D-NLS is rather intriguing, given that this G-actin variant does not have access to the cytoplasmic actin dynamics and is therefore unlikely to directly impinge on F-actin polymer turnover.

In sum, the principal mechanism regulating activity of the actin-MRTF–SRF connection appears highly conserved between different cell types, although – as exemplified for neurons – variations on a common theme might exist.

2   Essential functions of SRF and MRTF in Mouse Brain Development

2.1   Embryonic Neuronal Migration

Many neuronal phenotypes evoked by SRF deficiency point to an intimate link between neuronal motility and SRF-directed actin dynamics (Alberti et al. 2005; Knoll and Nordheim 2009; Knoll 2010). Constitutive *Srf* deletion mutant embryos die at the beginning of gastrulation (E6.5), precluding analysis of SRF functions in the nervous system (Arsenian et al. 1998). To circumvent this limitation, conditional *Srf* mutant mice were analyzed by various laboratories (Wiebel et al. 2002; Alberti et al. 2005; Ramanan et al. 2005; Etkin et al. 2006). We used a CamKIIα-iCre mouse strain (Casanova et al. 2001), allowing for neuronal SRF ablation commencing at late prenatal and early postnatal stages (Alberti et al. 2005). These $Srf^{flex1::CamKII\alpha\text{-}iCre}$ mice revealed a severe phenotype, including weight loss, ataxia and death at around three weeks after birth (Alberti et al. 2005).

An early step of nervous system development includes neuronal cell migration resulting in formation of distinct brain layers (Valiente and Marin 2010). In the brain, cells migrate by two different modes: tangential and radial migration (Ayala et al. 2007). Conditional forebrain-specific SRF ablation using the $Srf^{flex1::CamKII\alpha\text{-}iCre}$ mice revealed that SRF contributes to both types of neuronal cell migration. Firstly, impaired tangential cell migration along the rostral migratory stream was observed upon SRF deletion. Instead, progenitor cells failed to migrate and populate the olfactory bulb and ectopically accumulated in their area of origin, the subventricular zone (Alberti et al. 2005). Within these accumulating SRF-deficient cells, F-actin density was significantly reduced and inhibitory phosphorylation of the F-actin severing protein cofilin was drastically elevated (Alberti et al. 2005). Moreover, granule cells of the hippocampus were organized in a dispersed fashion, pointing to a role for SRF in radial cell migration (Stritt and Knoll 2010). Here SRF appears to be linked to reelin signaling (Stritt and Knoll 2010), a major signaling entity in neuronal cell migration (Forster et al. 2006).

Recent data obtained with conditional *Mrtfa/Mrtfb* double mutant mice strongly argue that, in neurons, SRF teams up with MRTFs to modulate neuronal cell migration (Mokalled et al. 2010). Upon *Mrtfa/Mrtfb* ablation, neuronal cell migration along the rostral migratory stream was affected to a comparable extent as previously observed upon conditional *Srf* mutagenesis (Alberti et al. 2005). Also, the enhanced P-cofilin levels previously seen with SRF-deficient neuronal cells were found in MRTF-deficient neurons (Mokalled et al. 2010). This finding was correlated with impaired regulation of the cofilin phosphorylating kinase LimK, due to downregulation of the Cdk5-associated kinase Pctaire (Mokalled et al. 2010). Interestingly, downregulation of Cdk5 and its co-regulator, p35, was previously observed in SRF-deficient neurons (Etkin et al. 2006). Therefore, inhibition of cofilin activity in cells deficient for either SRF or MRTF is due to impaired

Pctaire/Cdk5 activity, indirectly leading to elevated LimK activity. This regulatory loop nicely demonstrates how impaired transcription factor activity can lead to altered actin dynamics.

2.2   Axonal Outgrowth, Guidance, and Synaptic Targeting

Subsequent to cell migration, neurons elaborate neurites differentiating into axons and dendrites, a process termed neuronal polarization. In *Srf* mutants, neurite outgrowth, axon vs. dendrite differentiation and axon guidance of hippocampal and sensory axons are impaired (Knoll et al. 2006; Wickramasinghe et al. 2008; Stern et al. 2009). In addition, formation of a stereotypic dendritic arborization pattern and postsynaptic spines of CA1 pyramidal neurons is regulated by SRF (Stritt and Knoll 2010).

In vitro, SRF-deficient neurons are bipolar in shape and reduced in neurite length, and growth cones do not protrude finger-like filopodia, causing instead a rounded cell morphology (Knoll et al. 2006; Stern et al. 2009). Stimulation of growth cone cytoskeletal dynamics with axon guidance cues, such as ephrins (Knoll and Drescher 2002), results in a transient break-down of the cytoskeleton, the so-called growth cone collapse (Dent and Gertler 2003; Knoll et al. 2006; Pak et al. 2008; Pandithage et al. 2008). The growth cone collapse is thought to allow for adjusting axonal growth direction, thereby ensuring axonal target guidance in vivo. In *Srf* mutant neurons, cytoskeletal rearrangements fail to be elicited upon ephrin stimulation. Instead, aberrant growth cone structures consisting of F-actin and microtubule "rings" persisted (Knoll et al. 2006; Stern et al. 2009), which might be due to impaired activity of the actin-severing protein cofilin upon SRF deficiency (Alberti et al. 2005) that might impair proper F-actin turnover. It has not yet been demonstrated unambiguously that ephrin signaling via Eph receptor tyrosine kinases modulates SRF transcriptional responses to ensure a growth cone collapse. Alternatively, it is possible that, in SRF-deficient neurons, the growth cone-associated actin cytoskeleton is altered structurally in such way that any exogenous guidance cue fails to exert its normal impact on actin dynamics. In vitro treatments of SRF-deficient neurons with semaphorins (Knoll et al. 2006) and reelin (Stritt and Knoll 2010) also fail to exert their usual impact on neuronal morphology.

So far, the neurotrophins brain-derived neurotrophic factor (BDNF) and nerve growth factor (NGF) have emerged as true signaling growth factors, augmenting SRF-mediated transcriptional target gene expression (Stritt and Knoll 2010; Kalita et al. 2006; Wickramasinghe et al. 2008; Knoll and Nordheim 2009). Further, reelin (Stritt and Knoll 2010) and TGFß (Stritt et al. 2009) signaling is connected to SRF and can enhance neuronal gene expression in an SRF-dependent manner.

In vivo, SRF deficiency resulted in aberrant guidance of hippocampal mossy fibers (Knoll et al. 2006) and axons of the peripheral nervous system (Wickrama-singhe et al. 2008). In the mutant hippocampus, mossy fiber axons ectopically entered the cell body region of CA3 pyramidal neurons (stratum pyramidale) rather than navigating outside in the so-called stratum lucidum (Knoll et al. 2006). This finding might suggest impairment of repulsive axon guidance mechanisms by SRF ablation, thereby allowing for unrestrained ingrowth of axons inside the stratum pyramidale. Alternatively, this misrouting of mossy fibers might be explained by attractive or adhesive axon guidance forces that are up-regulated by *Srf* mutagenesis. In any case, axon path-finding errors in *Srf* mutants also affected synaptic targeting, as mossy fibers did not terminate blindly. Instead, ectopic mossy fiber boutons were found at CA3 pyramidal neuron somata or somatic protrusions rather than at their usual location, the CA3 dendrites (Knoll et al. 2006).

3   Functional Actin-SRF Interplay in the Mature Brain

3.1   Activity-Induced Gene Expression

In addition to the *Srf* deletion phenotypes observed in early postnatal brain development, conditional *Srf* mutagenesis also revealed essential SRF functions in the adult brain (for review, see Knoll and Nordheim 2009; Knoll 2010). These functions include neuronal activity-induced gene expression, synaptic plasticity, learning and memory (Dash et al. 2005; Ramanan et al. 2005; Etkin et al. 2006; Kalita et al. 2006; Lindecke et al. 2006; Nikitin and Kozyrev 2007; Tyan et al. 2008; Pintchovski et al. 2009).

Classical experimental paradigms for activity-induced gene expression include responses elicited upon (voluntary) exploration of a novel environment or upon (forced) induction of neuronal activity by electroconvulsive shocks. In such conditions, coordinated transcriptional activation of immediate-early genes (IEG; e.g. c-*fos*, c-Jun, Fos-B, Egr-1, *Arc*) is dependent upon SRF (Greenberg et al. 1985; Curran and Morgan 1995; Ramanan et al. 2005; Kawashima et al. 2009; Pintchovski et al. 2009). Ca^{2+} signaling contributes in important ways to this gene response (Bading et al. 1993; Miranti et al. 1995). Which individual SRF cofactors, i.e., which members of the TCF or MRTF family, mediate these responses in each of the different physiological circumstances remains to be clarified.

3.2   Learning and Memory

Long-term memory consolidation apparently requires de novo protein synthesis (Davis and Squire 1984). In line with SRF being required for activity-induced gene expression, neuronal *Srf* knock-out mutagenesis leads to impairment of long-term synaptic potentiation (LTP), either mildly (Etkin et al. 2006) or strongly (Ramanan et al. 2005). Besides impaired LTP, long-term synaptic depression (LTD) was obstructed upon SRF depletion (Etkin et al. 2006). This finding correlated with impaired formation of a hippocampus-dependent immediate memory of a novel context, resulting in, for example, decreased habituation during exploration of novel open field environments (Etkin et al. 2006). Recently, further evidence of a role for MRTF–SRF signaling in LTD was provided (Smith-Hicks et al. 2010). The late phase of LTD was blocked in mouse-cultured cerebellar Purkinje cells upon inhibition of either SRF or MRTF, whereas the early phase of LTD does not seem to require SRF. In agreement, deletion of the SRF target gene *Arc* impaired late phase LTD. The function of Arc in eliciting LTD is dependent upon the binding of SRF to an *Arc* gene regulatory element positioned 6.9 kb upstream of the gene (Smith-Hicks et al. 2010). Collectively, these studies strongly suggest that LTD requires SRF. The associated involvement of MRTF is indicated, although not finally proven, and it remains to be uncovered which role actin dynamics might be contributing to this response. Interestingly, structural alterations of synapses require actin dynamics (Cingolani and Goda 2008). It seems reasonable that structural rearrangements in the synaptic actin scaffold are conveyed through the actin-MRTF–SRF module. Notably, learning-associated alterations of neuronal MRTF expression in the hippocampus were recently reported (O'Sullivan et al. 2009).

4 Linking SRF to Neurodegenerative Diseases, Psychiatric Disorders, and Addiction

Based on the functions regulated by the actin-MRTF–SRF signaling entity in developmental and adult nervous system physiology (see above), this pathway is a potential contributor to neuronal pathology.

Our knowledge about the pathological implications of SRF is expanding continuously, covering multiple neurological disorders, including de-myelination, axonal degeneration, epilepsy, Alzheimer's disease (AD), addiction, and hyperactivity syndrome.

4.1 SRF Protects Against Axonal De-myelination and Degeneration

Until now, SRF had not been implicated as a major regulator of neuronal survival/ apoptosis in vivo. Thus, major neuron losses in peripheral and central brain regions have not been reported in various conditional SRF-deficient mice (Alberti et al. 2005; Ramanan et al. 2005; Etkin et al. 2006; Knoll et al. 2006; Wickramasinghe et al. 2008). Rather, in cell culture, neuro-protective functions have been attributed to SRF (Chang et al. 2004).

Nevertheless, in conditional, neuron-restricted SRF ablation in newborn mice, we found accumulation of markers frequently connected to axonal degeneration, although at two weeks after birth total axon number was unaltered (Stritt et al. 2009). These findings included microglia invasion, astrogliosis and enhanced expression of non-phosphorylated neurofilaments and amyloid precursor protein (APP) in the *corpus callosum*. This potential axonal degeneration was accompanied by de-myelination of *corpus callosal* axons. SRF regulates differentiation of myelin-producing oligodendrocytes via a paracrine mechanism that involves SRF-mediated transcriptional regulation of insulin growth factor signaling in neurons, which – upon secretion from neurons – affects oligodendrocytes in a paracrine manner.

These data suggest a neuroprotective function for SRF, preserving both the axon and surrounding myelin sheath. Nowadays multiple sclerosis (MS), resulting in myelin and axon loss, is considered to be based on alterations in the axonal compartment (Siffrin et al. 2010). Thus, the reported neuron-specific SRF conditional mouse mutant might be an interesting MS model system to study the consequences of axonal dysfunction on subsequent myelin loss.

4.2   SRF and Epileptic Seizure-Induced Neuronal IEG Expression

SRF is one of the central regulators of a physiological IEG response in neurons upon synaptic activity (Ramanan et al. 2005). Repetitive, seizure-induced synaptic transmission likewise results in rapid induction of IEG mRNAs (e.g., *Arc, c-fos, Egr1*) in the epileptic hippocampus (Morgan et al. 1987; Herdegen and Leah 1998; Cesari et al. 2004; Rakhade and Jensen 2009). As SRF is crucially involved in mediating a physiological IEG response in neurons, it is tempting to speculate that SRF might likewise emerge as an important regulator in conveying IEG induction upon seizures. In mouse mutants of the TCF family of SRF cofactors, IEG induction in kainate-induced epileptic seizures was only mildly impaired (Cesari et al. 2004), suggesting that the different TCFs might either be functionally redundant or even dispensable, which in turn suggests the possibility that SRF recruitment of the MRTF cofactors contributes to conveying a seizure-induced IEG response.

Interestingly, SRF protein itself, phospho-SRF levels (Herdegen et al. 1997), and SRF DNA affinity (Morris et al. 1999) are upregulated in the epileptic rodent brain. Yet, currently it remains unclear whether this potential increase in SRF transcriptional activity is beneficial (i.e., neuroprotective) to neurons or instead further enhances detrimental seizure effects.

4.3   SRF in M. Alzheimer

SRF and MRTFs are important partners regulating contractility in all types of muscle cell (for review, see Olson and Nordheim 2010). In addition, SRF–MRTF can modulate the blood flow through vessels via regulation of contractility genes and smooth muscle cell differentiation (Pipes et al. 2006; Miano et al. 2007). The latter SRF–MRTF function emerges as a novel therapeutic access to cure cognitive decline associated with reductions in cerebral blood flow (CBF) and dysregulation of cerebral vessel function contributing to AD (Chow et al. 2007). SRF and myocardin are induced in smooth muscle cells of AD patients (Chow et al. 2007). SRF and myocardin also promote a hypercontractile smooth muscle phenotype in arteries of AD models through expression of contractile genes (Chow et al. 2007). Together, the above mentioned changes in smooth muscle cell function might result in reduced CBF and contribute to neurovascular dysfunction and AD-associated dementia. In addition, SRF and myocardin contribute to AD via regulation of Aß clearance (Bell et al. 2009). SRF-myocardin over-expression suppresses expression of an Aß clearance receptor (LRP), thereby promoting focal Aβ brain accumulations in vessels (Bell et al. 2008).

4.4 SRF as a Novel Player in Addictive Behavior and Hyperactive Disorder

As described above, SRF conveys activity-dependent gene induction triggered by a wealth of stimuli, including physiological neuronal transmission and forced (e.g., electro-convulsive shocks, seizures; see above) and voluntary synaptic activity, as achieved by exploration in a novel enriched environment (Herdegen et al. 1997; Ramanan et al. 2005; Etkin et al. 2006). Similarly, SRF was identified as an important regulator to convey transcription of a subset of cocaine-induced genes in dopaminoceptive neurons (Parkitna et al. 2010). Further, it was found that loss of SRF in dopaminoceptive, but not dopaminergic, neurons is responsible for the development of a hyperactivity syndrome and deficits in habituation processes. This phenotype resembles children suffering from attention deficit hyperactivity disorder (ADHD). Thus, it will be interesting to explore whether polymorphisms in the *Srf* gene or in loci of SRF cofactors are associated with ADHD.

In line with these results, a constitutively active SRF protein (SRF-VP16) was recently attributed a protective function in a rodent model of fetal alcohol spectrum disorders (FASD; Paul et al. 2010). The underlying mechanism by which SRF restores ocular dominance synaptic plasticity in alcohol-treated animals has not yet been identified. However, similar to SRF's impact on oligodendrocytes (see above; Stritt et al. 2009), SRF appears to operate at least in part via a paracrine mechanism.

Actin dynamics and behavioral responses by excessive alcohol consumption are tightly connected on a molecular and cellular level (Offenhauser et al. 2006). Thus, the actin-MRTF–SRF module might emerge as a central signaling cascade targeted in addictive and cognitive disorders.

Therefore, as exemplified by the various neuropathological conditions outlined above, interfering with SRF activity in brain pathology might be a double-edged sword, potentially resulting in either beneficial or detrimental effects.

Acknowledgments AN acknowledges financial support from the DFG through grant NO 120/12-3. BK is supported by the DFG Emmy Noether program, the Hertie Foundation, and the Schram Foundation.

References

Alberti S, Krause SM, Kretz O, Philippar U, Lemberger T, Casanova E, Wiebel FF, Schwarz H, Frotscher M, Schutz G, Nordheim A (2005) Neuronal migration in the murine rostral migratory stream requires serum response factor. Proc Natl Acad Sci USA 102:6148–6153

Arsenian S, Weinhold B, Oelgeschlager M, Ruther U, Nordheim A (1998) Serum response factor is essential for mesoderm formation during mouse embryogenesis. Embo J 17:6289–6299

Ayala R, Shu T, Tsai LH (2007) Trekking across the brain: the journey of neuronal migration. Cell 128:29–43

Bading H, Ginty DD, Greenberg ME (1993) Regulation of gene expression in hippocampal neurons by distinct calcium signaling pathways. Science 260:181–186

Barnes AP, Polleux F (2009) Establishment of axon-dendrite polarity in developing neurons. Annu Rev Neurosci 32:347–381

Bell RD, Deane R, Chow N, Long X, Sagare A, Singh I, Streb JW, Guo H, Rubio A, Van Nostrand W, Miano JM, Zlokovic BV (2009) SRF and myocardin regulate LRP-mediated amyloid-beta clearance in brain vascular cells. Nat Cell Biol 11:143–153

Buchwalter G, Gross C, Wasylyk B (2004) Ets ternary complex transcription factors. Gene 324:1–14

Casanova E, Fehsenfeld S, Mantamadiotis T, Lemberger T, Greiner E, Stewart AF, Schutz G (2001) A CamKIIalpha iCre BAC allows brain-specific gene inactivation. Genesis 31:37–42

Cesari F, Brecht S, Vintersten K, Vuong LG, Hofmann M, Klingel K, Schnorr JJ, Arsenian S, Schild H, Herdegen T, Wiebel FF, Nordheim A (2004) Mice deficient for the ets transcription factor elk-1 show normal immune responses and mildly impaired neuronal gene activation. Mol Cell Biol 24:294–305

Chang SH, Poser S, Xia Z (2004) A novel role for serum response factor in neuronal survival. J Neurosci 24:2277–2285

Chow N, Bell RD, Deane R, Streb JW, Chen J, Brooks A, Van Nostrand W, Miano JM, Zlokovic BV (2007) Serum response factor and myocardin mediate arterial hypercontractility and cerebral blood flow dysregulation in Alzheimer's phenotype. Proc Natl Acad Sci USA 104:823–828

Cingolani LA, Goda Y (2008) Actin in action: the interplay between the actin cytoskeleton and synaptic efficacy. Nat Rev Neurosci 9:344–356

Curran T, Morgan JI (1995) Fos: an immediate-early transcription factor in neurons. J Neurobiol 26:403–412

Dash PK, Orsi SA, Moore AN (2005) Sequestration of serum response factor in the hippocampus impairs long-term spatial memory. J Neurochem 93:269–278

Davis HP, Squire LR (1984) Protein synthesis and memory: a review. Psychol Bull 96:518–59

Dent EW, Gertler FB (2003) Cytoskeletal dynamics and transport in growth cone motility and axon guidance. Neuron 40:209–227

Etkin A, Alarcon JM, Weisberg SP, Touzani K, Huang YY, Nordheim A, Kandel ER (2006) A role in learning for SRF: deletion in the adult forebrain disrupts LTD and the formation of an immediate memory of a novel context. Neuron 50:127–143

Forster E, Zhao S, Frotscher M (2006) Laminating the hippocampus. Nat Rev Neurosci 7:259–267

Greenberg ME, Greene LA, Ziff EB (1985) Nerve growth factor and epidermal growth factor induce rapid transient changes in proto-oncogene transcription in PC12 cells. J Biol Chem 260:14101–14110

Herdegen T, Leah JD (1998) Inducible and constitutive transcription factors in the mammalian nervous system: control of gene expression by Jun, Fos and Krox, and CREB/ATF proteins. Brain Res Brain Res Rev 28:370–490

Herdegen T, Blume A, Buschmann T, Georgakopoulos E, Winter C, Schmid W, Hsieh TF, Zimmermann M, Gass P (1997) Expression of activating transcription factor-2, serum response factor and cAMP/Ca response element binding protein in the adult rat brain following generalized seizures, nerve fibre lesion and ultraviolet irradiation. Neuroscience 81:199–212

Kalita K, Kharebava G, Zheng JJ, Hetman M (2006) Role of megakaryoblastic acute leukemia-1 in ERK1/2-dependent stimulation of serum response factor-driven transcription by BDNF or increased synaptic activity. J Neurosci 26:10020–10032

Kawashima T, Okuno H, Nonaka M, Adachi-Morishima A, Kyo N, Okamura M, Takemoto-Kimura S, Worley PF, Bito H (2009) Synaptic activity-responsive element in the Arc/Arg3.1 promoter essential for synapse-to-nucleus signaling in activated neurons. Proc Natl Acad Sci USA 106:316–321

Knoll B (2010) Actin-mediated gene expression in neurons: the MRTF-SRF connection. Biol Chem 391:591–597

Knoll B, Drescher U (2002) Ephrin-As as receptors in topographic projections. Trends Neurosci 25:145–149

Knoll B, Nordheim A (2009) Functional versatility of transcription factors in the nervous system: the SRF paradigm. Trends Neurosci 32:432–442

Knoll B, Kretz O, Fiedler C, Alberti S, Schutz G, Frotscher M, Nordheim A (2006) Serum response factor controls neuronal circuit assembly in the hippocampus. Nat Neurosci 9:195–204

Lindecke A, Korte M, Zagrebelsky M, Horejschi V, Elvers M, Widera D, Prullage M, Pfeiffer J, Kaltschmidt B, Kaltschmidt C (2006) Long-term depression activates transcription of imme-diate early transcription factor genes: involvement of serum response factor/Elk-1. Eur J Neurosci 24:555–563

Medjkane S, Perez-Sanchez C, Gaggioli C, Sahai E, Treisman R (2009) Myocardin-related transcription factors and SRF are required for cytoskeletal dynamics and experimental metas-tasis. Nat Cell Biol 11:257–268

Miano JM, Long X, Fujiwara K (2007) Serum response factor: master regulator of the actin cytoskeleton and contractile apparatus. Am J Physiol Cell Physiol 292:C70–81

Miralles F, Posern G, Zaromytidou AI, Treisman R (2003) Actin dynamics control SRF activity by regulation of its coactivator MAL. Cell 113:329–342

Miranti CK, Ginty DD, Huang G, Chatila T, Greenberg ME (1995) Calcium activates serum response factor-dependent transcription by a Ras- and Elk-1-independent mechanism that involves a Ca2+/calmodulin-dependent kinase. Mol Cell Biol 15:3672–3684

Mokalled MH, Johnson A, Kim Y, Oh J, Olson EN (2010) Myocardin-related transcription factors regulate the Cdk5/Pctaire1 kinase cascade to control neurite outgrowth, neuronal migration and brain development. Development 137:2365–2374

Morgan JI, Cohen DR, Hempstead JL, Curran T (1987) Mapping patterns of c-fos expression in the central nervous system after seizure. Science 237:192–197

Morris TA, Jafari N, Rice AC, Vasconcelos O, DeLorenzo RJ (1999) Persistent increased DNA-binding and expression of serum response factor occur with epilepsy-associated long-term plasticity changes. J Neurosci 19:8234–8243

Nikitin VP, Kozyrev SA (2007) Transcription factor serum response factor is selectively involved in the mechanisms of long-term synapse-specific plasticity. Neurosci Behav Physiol 37:83–88

Offenhauser N, Castelletti D, Mapelli L, Soppo BE, Regondi MC, Rossi P, D'Angelo E, Frassoni C, Amadeo A, Tocchetti A, Pozzi B, Disanza A, Guarnieri D, Betsholtz C, Scita G, Heberlein U, Di Fiore PP (2006) Increased ethanol resistance and consumption in Eps8 knockout mice correlates with altered actin dynamics. Cell 127:213–226

Olson EN, Nordheim A (2010) Linking actin dynamics and gene transcription to drive cellular motile functions. Nat Rev Mol Cell Biol 11:353–365

O'Sullivan NC, Pickering M, Di Giacomo D, Loscher JS, Murphy KJ (2009) Mkl transcription cofactors regulate structural plasticity in hippocampal neurons. Cereb Cortex 20:1915–1925

Pak CW, Flynn KC, Bamburg JR (2008) Actin-binding proteins take the reins in growth cones. Nat Rev Neurosci 9:136–147

Pandithage R, Lilischkis R, Harting K, Wolf A, Jedamzik B, Luscher-Firzlaff J, Vervoorts J, Lasonder E, Kremmer E, Knoll B, Luscher B (2008) The regulation of SIRT2 function by cyclin-dependent kinases affects cell motility. J Cell Biol 180:915–929

Parkitna JR, Bilbao A, Rieker C, Engblom D, Piechota M, Nordheim A, Spanagel R, Schutz G (2010) Loss of the serum response factor in the dopamine system leads to hyperactivity. Faseb J 24:2427–2435

Paul AP, Pohl-Guimaraes F, Krahe TE, Filgueiras CC, Lantz CL, Colello RJ, Wang W, Medina AE (2010) Overexpression of serum response factor restores ocular dominance plasticity in a model of fetal alcohol spectrum disorders. J Neurosci 30:2513–2520

Pintchovski SA, Peebles CL, Kim HJ, Verdin E, Finkbeiner S (2009) The serum response factor and a putative novel transcription factor regulate expression of the immediate-early gene Arc/Arg3.1 in neurons. J Neurosci 29:1525–1537

Pipes GC, Creemers EE, Olson EN (2006) The myocardin family of transcriptional coactivators: versatile regulators of cell growth, migration, and myogenesis. Genes Dev 20:1545–1556

Posern G, Treisman R (2006) Actin' together: serum response factor, its cofactors and the link to signal transduction. Trends Cell Biol 16:588–596

Posern G, Sotiropoulos A, Treisman R (2002) Mutant actins demonstrate a role for unpolymerized actin in control of transcription by serum response factor. Mol Biol Cell 13:4167–4178

Posern G, Miralles F, Guettler S, Treisman R (2004) Mutant actins that stabilise F-actin use distinct mechanisms to activate the SRF coactivator MAL. Embo J 23:3973–3983

Rakhade SN, Jensen FE (2009) Epileptogenesis in the immature brain: emerging mechanisms. Nat Rev Neurol 5:380–391

Ramanan N, Shen Y, Sarsfield S, Lemberger T, Schutz G, Linden DJ, Ginty DD (2005) SRF mediates activity-induced gene expression and synaptic plasticity but not neuronal viability. Nat Neurosci 8:759–767

Schratt G, Philippar U, Berger J, Schwarz H, Heidenreich O, Nordheim A (2002) Serum response factor is crucial for actin cytoskeletal organization and focal adhesion assembly in embryonic stem cells. J Cell Biol 156:737–750

Siffrin V, Vogt J, Radbruch H, Nitsch R, Zipp F (2010) Multiple sclerosis – candidate mechanisms underlying CNS atrophy. Trends Neurosci 33:202–210

Smith-Hicks C, Xiao B, Deng R, Ji Y, Zhao X, Shepherd JD, Posern G, Kuhl D, Huganir RL, Ginty DD, Worley PF, Linden DJ (2010) SRF binding to SRE 6.9 in the Arc promoter is essential for LTD in cultured Purkinje cells. Nat Neurosci. doi:10.1038/nn.2611

Stern S, Debre E, Stritt C, Berger J, Posern G, Knoll B (2009) A nuclear actin function regulates neuronal motility by serum response factor-dependent gene transcription. J Neurosci 29:4512–4518

Stritt C, Knoll B (2010) Serum response factor regulates hippocampal lamination and dendrite development and is connected with reelin signaling. Mol Cell Biol 30:1828–1837

Stritt C, Stern S, Harting K, Manke T, Sinske D, Schwarz H, Vingron M, Nordheim A, Knoll B (2009) Paracrine control of oligodendrocyte differentiation by SRF-directed neuronal gene expression. Nat Neurosci 12:418–427

Tabuchi A, Estevez M, Henderson JA, Marx R, Shiota J, Nakano H, Baraban JM (2005) Nuclear translocation of the SRF co-activator MAL in cortical neurons: role of RhoA signalling. J Neurochem 94:169–180

Torreano PJ, Waterman-Storer CM, Cohan CS (2005) The effects of collapsing factors on F-actin content and microtubule distribution of Helisoma growth cones. Cell Motil Cytoskeleton 60:166–179

Tyan SW, Tsai MC, Lin CL, Ma YL, Lee EH (2008) Serum- and glucocorticoid-inducible kinase 1 enhances zif268 expression through the mediation of SRF and CREB1 associated with spatial memory formation. J Neurochem 105:820–832

Valiente M, Marin O (2010) Neuronal migration mechanisms in development and disease. Curr Opin Neurobiol 20:68–78

Vartiainen MK, Guettler S, Larijani B, Treisman R (2007) Nuclear actin regulates dynamic subcellular localization and activity of the SRF cofactor MAL. Science 316:1749–1752

Wickramasinghe SR, Alvania RS, Ramanan N, Wood JN, Mandai K, Ginty DD (2008) Serum response factor mediates NGF-dependent target innervation by embryonic drg sensory neurons. Neuron 58:532–545

Wiebel FF, Rennekampff V, Vintersten K, Nordheim A (2002) Generation of mice carrying conditional knockout alleles for the transcription factor SRF. Genesis 32:124–126

The E3 Ubiquitin Ligase Ube3A Regulates Synaptic Function Through the Ubiquitination of Arc

Eric C. Griffith, Paul L. Greer, and Michael E. Greenberg

Abstract The E3 ubiquitin ligase Ube3A was initially isolated as a cellular factor required for the transforming activity of human papillomavirus; however, subsequent studies have implicated this gene in aspects of human cognitive function. Maternal deficiency of Ube3A gives rise to Angelman syndrome (AS), a severe neurodevelopmental disorder, and dysregulation of this gene has also been implicated in autism. However, the specific roles that Ube3A plays in the developing and adult nervous system are largely unknown. We find that Ube3A is a neuronal activity-regulated protein that controls synaptic function by ubiquitinating and degrading the synaptic protein Arc. In the absence of Ube3A, elevated levels of Arc accumulate in neurons, resulting in the excessive internalization of AMPA-type glutamate receptors at synapses and impaired synaptic function. These findings suggest that excess Arc expression may contribute to at least some aspects of the cognitive dysfunction associated with AS.

UBE3A encodes an E3 ubiquitin ligase that has been shown to function as an enzymatic component of the ubiquitin-proteasome protein degradation pathway. Although initially identified as a cellular factor required for the transforming activity of human papillomavirus (Scheffner et al. 1993), subsequent genetic studies indicate a critical role for Ube3A in human cognitive function. Yet, the precise function of Ube3A during nervous system development and how Ube3A mutations give rise to cognitive impairment are not clear. After briefly reviewing the evidence implicating Ube3A in human cognitive function, we discuss recent findings from our laboratory concerning the role of Ube3A in regulating synaptic development and function (Greer et al. 2010). We find that experience-driven neuronal activity induces *Ube3a* transcription and that Ube3A then regulates excitatory synapse development by controlling the degradation of Arc, a synaptic protein that promotes the internalization of AMPA-type glutamate receptors. We

M.E. Greenberg (✉)
Department of Neurobiology, Harvard Medical School, 220 Longwood Avenue, Boston, MA 02115, USA
e-mail: meg@hms.harvard.edu

T. Curran and Y. Christen (eds.), *Two Faces of Evil: Cancer and Neurodegeneration*, Research and Perspectives in Alzheimer's Disease, DOI 10.1007/978-3-642-16602-0_4, © Springer-Verlag Berlin Heidelberg 2011

find that disruption of Ube3A function in neurons leads to an increase in Arc expression and a concomitant decrease in the number of AMPA receptors at excitatory synapses. We propose that this dysregulation of AMPA receptor trafficking may contribute to the cognitive dysfunction that occurs in Angelman syndrome (AS) and possibly autism spectrum disorders (ASDs).

1 A Role for Ube3A in Human Disorders of Cognitive Function

The first clue implicating Ube3A in nervous system development came from the observation that loss-of-function mutations in the *UBE3A* gene in humans give rise to AS, a debilitating neurodevelopmental disorder characterized by severe mental retardation, motor dysfunction, speech impairment, hyperactivity, and frequent seizures that affects approximately one in 15,000 individuals (Kishino et al. 1997; Matsuura et al. 1997). The *UBE3A* gene is subject to brain-specific genomic imprinting, with predominant transcription of the maternal allele in the brain. Consistent with this regulation, AS has also been found to be associated with de novo maternal deletions of chromosome 15q11-q13, paternal chromosome 15 uniparental disomy, and rare imprinting defects. Taken together, molecular genetic studies indicate that a failure to inherit a normal active maternal copy of the *UBE3A* gene accounts for 85–90% of all AS cases (Williams 2005). Given that the maternal imprinting of the *UBE3A* gene is restricted to specific brain regions, these findings support the idea that loss of Ube3A function in the central nervous system underlies AS pathology.

Beyond AS, genetic evidence also suggests a possible role for Ube3A dysregulation in the etiology of some non-syndromic ASDs. These complex, behaviorally defined disorders of early childhood are characterized by impaired sociability, language, and nonverbal communication skills, together with the presence of repetitive behaviors. Compelling evidence exists for a genetic basis for autism. Numerous studies, however, indicate that autism displays a high degree of genetic heterogeneity, and considerable effort has gone into identifying genetic variations that are associated with ASDs. In this regard, a growing body of evidence has implicated Ube3A dysregulation as a causative factor in cases of non-syndromic autistic disorders. First, AS-affected individuals exhibit a high prevalence of associated autism. Second, the *UBE3A* gene lies within human chromosome 15q11-13, one of the few regions of recurrent cytogenetic aberrations associated with autism. Indeed, it has been estimated that chromosomal duplications in this region occur in up to 5% of autism cases (Cook et al. 1997; Glessner et al. 2009; Sutcliffe et al. 2003; Veenstra-Vanderweele et al. 2004). In addition, given the restricted, maternal-specific expression of *UBE3A* in the brain, it is particularly striking that cases have been documented in which maternal, but not paternal, inheritance of 15q11-13 duplications is associated with autistic phenotypes (Cook et al. 1997).

In the case of AS, direct evidence for the causative role of Ube3A mutations has come from the analysis of mouse models generated by targeted inactivation of the *Ube3a* gene (Jiang et al. 1998; Miura et al. 2002). Consistent with the known brain-specific imprinting of the locus, inheritance of the mutant *Ube3a* allele through the maternal but not the paternal germline gives rise to several features of AS, including microcephaly, impaired motor function, and inducible seizures. Moreover, electrophysiological studies in these animals have revealed profound impairments in synaptic plasticity. *Ube3a* maternal-deficient ($Ube3a^{m-/p+}$) mice harbor deficits in cortical and hippocampal long-term potentiation (LTP) due to an alteration in the threshold of synaptic stimulation necessary for LTP induction (Weeber et al. 2003; Yashiro et al. 2009). Similarly, cortical long-term depression (LTD) is absent in young *Ube3a* maternal-deficient animals (Yashiro et al. 2009). Further analysis in the visual cortex has suggested that Ube3A is necessary for the maintenance of plasticity during experience-dependent synaptic remodeling (Sato and Stryker 2010; Yashiro et al. 2009). While normal neocortical plasticity was preserved in $Ube3a^{m-/p+}$ mice under conditions of sensory deprivation, plasticity was rapidly dampened by ongoing sensory experience in the absence of Ube3A. Remarkably, this loss of plasticity is reversible, as late-onset visual deprivation restored normal synaptic plasticity in *Ube3a* maternal-deficient animals. Consistent with these synaptic effects, $Ube3a^{m-/p+}$ mice also exhibit severe learning and memory deficits, including defects in spatial memory and context-dependent fear conditioning (Jiang et al. 1998; Miura et al. 2002).

The idea that Ube3A acts to modulate synaptic development is supported by the finding that the Ube3A protein is present in the pre- and post-synaptic compartments of cultured hippocampal neurons (Dindot et al. 2008). In addition, $Ube3a^{m-/p+}$ mice exhibit abnormal dendritic spine morphology indicative of altered synaptic connectivity (Dindot et al. 2008; Sato and Stryker 2010; Yashiro et al. 2009). Yet, despite these findings, the specific roles that Ube3A plays in the developing nervous system remain to be fully characterized. Indeed, to date, our understanding of the molecular functions of Ube3A stems primarily from its role as an E3 ubiquitin ligase. Ubiquitin ligases catalyze the terminal step of a series of reactions involving the transfer of the small polypeptide ubiquitin to the lysine residues of target proteins. Addition of a single ubiquitin molecule can act to modulate protein function, whereas multi-ubiquitination often leads to recognition by the 26S proteasome and protein degradation. In the case of Ube3A, there is a strong correlation between the AS-associated point mutations and the loss of E3 ubiquitin ligase activity (Cooper et al. 2004; Malzac et al. 1998), suggesting that dysregulation of neural Ube3A substrates underlies the development of AS. Consistent with this idea, a number of studies indicate an important role for ubiquitin-dependent proteolysis in the regulation of synapse formation and function (Patrick 2006). As of yet, however, the cellular targets of Ube3A activity remain largely unknown. Thus, despite the finding that mutation of Ube3A is the causative factor of AS, little is known about how the disruption of Ube3A activity results in cognitive impairment, and a treatment for this disorder remains elusive.

2   *Ube3a* is a Neuronal Activity-Regulated Gene

Current evidence indicates a role for Ube3A in experience-dependent synaptic remodeling. While neurotransmitter function at individual synapses clearly contributes to synaptic remodeling by mechanisms that do not require transcription, it is now clear that synaptic transmission leads to the activation of a transcriptional program in the postsynaptic cell that both positively and negatively regulates synapse development and plasticity (Greer and Greenberg 2008). Moreover, the initial emergence of AS and ASD symptoms during early postnatal development – the time during which activity-dependent genes are highly induced and experience-dependent synaptic remodeling is peaking – raises the possibility that Ube3A participates in this neuronal activity-dependent gene program. In support of this hypothesis, accumulating evidence indicates that mutations in a large number of the components of this activity-dependent gene program give rise to disorders of human cognition (Amir et al. 1999; Morrow et al. 2008; Petrij et al. 1995; Trivier et al. 1996; Zoghbi 2003).

To directly test whether Ube3A expression is responsive to neuronal activity, we assessed Ube3A mRNA and protein levels under various conditions of neuronal stimulation in culture and in vivo using quantitative RT–PCR and Western blotting (Greer et al. 2010). Membrane depolarization of cultured neurons using elevated levels of extracellular potassium to activate L-type voltage-sensitive calcium channels (L-VSCCs) led to a significant upregulation of neuronal *Ube3a* mRNA levels within 5 h of treatment. Glutamate or N-methyl-D-aspartate (NMDA) stimulation of cultured neurons also potentiated *Ube3a* mRNA levels, indicating that stimulation of ligand-gated ion channels that mediate excitatory neurotransmission can also trigger Ube3A expression. Conversely, blocking endogenous neuronal activity using glutamate receptor and/or sodium channel antagonists resulted in a significant decrease in the levels of *Ube3a* mRNA. For all tested conditions, Ube3A protein levels were found to mirror the observed changes in mRNA abundance.

To determine whether Ube3A levels in the brains of intact animals are responsive to synaptic activity, we first asked if synchronized synaptic activity induced by chemoconvulsants acutely affects Ube3A expression in the hippocampus. Indeed, seizures induced by intraperitoneal injection of kainic acid were found to result in a significant increase in Ube3A expression at both the mRNA and protein level as compared to saline-injected controls. In addition, we also examined the response of Ube3A levels to natural modes of synaptic stimulation. Exploration of a novel environment was found to lead to a robust upregulation of Ube3A mRNA and protein levels in the mouse hippocampus as compared to levels in control littermates exposed to a standard laboratory cage (Fig. 1a and data not shown). Taken together, these results demonstrate that *Ube3a* levels are regulated by synaptic activity in the intact brain, raising the possibility that the absence of experience-driven *Ube3a* induction may contribute to the neurological impairment in AS.

In humans and rodents, the *Ube3a* gene encompasses a complex genomic locus, with sequence data indicating a variety of mRNA transcripts that are likely transcribed from at least three distinct promoters. Of these *Ube3a* transcripts, those

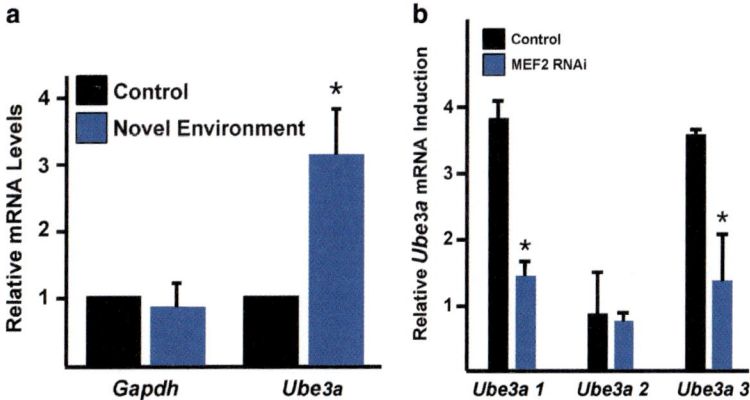

Fig. 1 Regulation of Ube3A expression by neuronal activity and MEF2 transcription factors. (**a**) Quantitative RT-PCR analysis of *Ube3a* and *Gapdh* mRNA from 6-week-old CD1 male mice placed in standard laboratory cages (control) or in cages containing novel objects (enriched environment). After 3 h, mice were sacrificed and RNA was extracted. Data are presented as mean ±SEM from three independent experiments. *Asterisk* indicates statistical significance in pairwise comparison (P< 0.05, T-test). (**b**) Quantitative RT-PCR analysis of the three *Ube3a* transcripts from mRNA extracted from hippocampal neurons transduced with lentivirus expressing either control shRNA or shRNAs targeting MEF2A and MEF2D. Neurons were stimulated with 55 mM potassium chloride (KCl) for 6 h before RNA was harvested. Data are plotted as fold induction in stimulated cells over unstimulated cells and are presented as mean ±SEM from three independent experiments. *Asterisk* indicates statistical significance in pairwise comparison (P< 0.01, T-test). Adapted from Greer et al. (2010) with permission

initiated by promoters 1 and 3 are induced in response to neuronal activity. Sequence analysis of these proximal promoter regions revealed the presence of evolutionarily conserved consensus binding sites for the myocyte enhancer factor 2 (MEF2) family of activity-regulated transcription factors. Chromatin immuno-precipitation studies confirmed binding of MEF2 to these *cis*-acting regulatory elements. Moreover, the neuronal activity-dependent induction of *Ube3a* promoter 1- and 3-driven mRNA transcripts is significantly reduced in neurons infected with lentiviruses encoding short hairpin RNAs (shRNAs) that specifically target MEF2, but it is not affected by the presence of control shRNAs (Fig. 1b). By contrast, the expression of *Ube3a* promoter 2-dependent mRNA transcripts, as well as the *Gapdh*, *β3-tubulin*, and *Crem* mRNAs, is unaffected by the presence of MEF2 shRNA. Thus, our findings indicate that, in response to neuronal activity, *Ube3a* mRNA and protein expression are induced by a MEF2-dependent mechanism.

3 Regulation of Synaptic AMPA Receptor Function by Ube3A

The presence of potential MEF2-binding sites within *Ube3a* regulatory regions was of interest because MEF2 proteins have been shown to modulate synaptic development and regulate genes implicated in ASDs (Flavell et al. 2006, 2008; Morrow et al. 2008). As neuronal activity via MEF2 plays a crucial role in regulating excitatory synapse development, we further investigated the role of Ube3A in regulating synaptic development and function. We first examined the effect of reducing neuronal Ube3A levels on miniature excitatory postsynaptic currents (mEPSCs), which have proven to be a sensitive measure of neuronal connectivity. To this end, we acutely reduced Ube3A expression in hippocampal neurons using shRNAs that specifically target Ube3A, and we recorded mEPSCs from transfected neurons 4 days after transfection using a whole-cell voltage clamp configuration in the presence of tetrodotoxin (TTX) and bicuculline. Whereas no effect on mEPSC amplitude, input resistance, or whole cell capacitance was observed following *Ube3a* knockdown, we did observe a significant decrease in mEPSC frequency in *Ube3a* shRNA-expressing cells. The frequency of mEPSCs is a function of the number of α-amino-3-hydroxy-5-methyl-4-isoxazolepropionic acid-type glutamate receptor (AMPAR)-containing synaptic connections as well as the presynaptic probability of release. Because our experimental paradigm involved the postsynaptic suppression of Ube3A expression in the context of low-density cultures, we first considered possible postsynaptic changes that might account for the observed change in mEPSC frequency.

To examine the number of excitatory synapses formed onto transfected neurons, cells were fixed and stained for the presynaptic marker synapsin-1, as well as the postsynaptic marker postsynaptic density protein 95 (PSD-95). Quantification of opposed synapsin-1/PSD-95 puncta on control and shRNA-expressing cells by confocal microscopy revealed no effect of Ube3A knockdown on co-cluster density over the time course of the experiment. Consistent with this finding, four days after transfection no difference in dendritic spine density was observed in Ube3A shRNA-expressing neurons as compared to control transfected neurons. These tiny protrusions from the surface of dendrites, with a thin neck and expanded head, act as sites of signal integration and plasticity (Harris and Kater 1994), and, in the hippocampus, each spine head typically houses a single excitatory synapse. Thus, we conclude that Ube3A loss does not affect excitatory synapse number within the short period of time needed to observe alterations in mEPSC frequency.

We next examined the possibility that the observed effect of Ube3A knockdown on mEPSC frequency might reflect alterations in postsynaptic AMPAR expression. In this regard, expression of either of two shRNAs targeting Ube3A was found to cause an acute reduction in the levels of surface AMPARs expressed at the synapse within this short time period. For these studies, we focused our attention on the GluR1 subunit of the AMPA receptor as GluR1 insertion into the plasma membrane is known to be regulated by neuronal activity (Greger and Esteban 2007), staining hippocampal neurons with anti-GluR1 antibodies under non-permeablizing conditions and quantifying the number of GluR1 puncta that co-localize with the synaptic marker PSD-95. The observed decrease in surface GluR1 could be rescued by co-expression of an RNAi-resistant form of Ube3A and was not due to a change in the synthesis or degradation of AMPARs, as wild-type and Ube3A-deficient cells expressed equal amounts of GluR1 and GluR2 subunits. Moreover, this effect was selective insofar as cell surface levels of the NR1 subunit of the NMDA-type glutamate receptor (NMDAR) were unaltered in response to Ube3A knockdown.

To determine whether this effect reflected altered AMPAR endocytosis in the absence of Ube3A, we used GluR1-specific antibodies to label surface AMPARs on neurons transfected with shRNAs targeting Ube3A. Membrane depolarization was used to induce the endocytosis of synaptic AMPARs, and acid stripping was used to remove remaining antibodies from the cell surface. Subsequent permeablization of the cells and staining with fluorescent secondary antibodies to detect newly internalized GluR1 revealed increased levels of endocytosed GluR1 in Ube3A shRNA-expressing neurons compared to control transfected cells. Thus, the decreased synaptic expression of AMPARs in Ube3A knockdown cells is due, at least in part, to an increase in AMPAR endocytosis. Although we cannot rule out the possibility that a decrease in Ube3A expression affected the presynaptic probability of release, we believe the observed increase in AMPAR endocytosis favors the hypothesis that the decrease in mEPSC frequency following Ube3A knockdown reflects the loss of AMPARs from a subset of synapses.

These findings suggest that, in AS, the absence of Ube3A activity may result in the reduction in the expression of AMPARs at synapses. To investigate this possibility, we examined AMPAR expression and function in a *Ube3a* mutant mouse strain that recapitulates features of AS (Jiang et al. 1998). We first employed array tomography, a method in which ultra-thin sections of brain tissue are stained, imaged, and synapses visualized as a three-dimensional reconstruction (Micheva and Smith 2007), to determine whether synaptic GluR1 expression is dysregulated in $Ube3a^{m-/p+}$ neurons in the context of an intact neural circuit. Using this technique, we observed a significant decrease in GluR1 synaptic localization in the hippocampi of $Ube3a^{m-/p+}$ mice. In contrast, staining for the NR1 subunit of the NMDA receptor revealed no differences in synaptic expression between wild-type and *Ube3a* maternal-deficient animals, suggesting that the synaptic expression of AMPARs is selectively decreased in $Ube3a^{m-/p+}$ mice. Importantly, this decrease in synaptic AMPAR expression in $Ube3a^{m-/p+}$ mice is not the result of an overall decrease in GluR1 expression, as hippocampi from wild-type and *Ube3a* maternal-deficient mice express similar levels of GluR1.

To determine whether the decreased expression of AMPARs at the synapses of $Ube3a^{m-/p+}$ mice is associated with a functional decrease in synaptic transmission in these animals, whole-cell recordings were made from CA1 pyramidal neurons, and the resultant AMPAR- and NMDAR-mediated currents were measured to assess glutamate receptor function at individual synapses. Whereas NMDAR currents were unaffected in Ube3A-deficient samples, we observed a significant decrease in AMPAR-mediated currents in $Ube3a^{m-/p+}$ mice as compared to those in wild-type mice, leading to a shift in the ratio of synaptic AMPAR/NMDAR-mediated currents. Taken together, these observations support the conclusion that AMPAR expression and function at synapses are significantly affected in *Ube3a* maternal-deficient animals.

4 A Novel Proteomic Approach to Ube3A Substrate Identification

Since Ube3A is a ubiquitin ligase and the majority of the mutations isolated from individuals with AS abrogate Ube3A ubiquitin ligase activity (Cooper et al. 2004), we hypothesized that one way in which Ube3A might act to regulate synaptic development is by ubiquitinating and degrading proteins that are important for AMPAR endocytosis and/or synapse development or maintenance. To begin to test this possibility, we generated a dominant interfering mutant form of Ube3A in which a catalytic cysteine residue is mutated to an alanine (Ube3A C833A; Kumar et al. 1999). Indeed, overexpression of Ube3A C833A, but not wild-type Ube3A, led to a significant reduction in cell surface AMPAR on transfected neurons, consistent with the idea that Ube3A's ubiquitin ligase activity is crucial for its effects on synaptic function. As of yet, however, the cellular targets of Ube3A's E3 ubiquitin ligase activity remain largely unknown, and those few Ube3A substrates that have been identified provide little insight into AS etiology. Thus, we sought an approach by which to identify novel neuronal substrates of Ube3A.

Ubiquitin ligases catalyze the terminal step of a series of reactions involving the transfer of the small polypeptide ubiquitin to the lysine residues of target proteins. The identification of the substrates for particular E3 ligases has generally proven challenging. One potential approach involves the global analysis of ubiquitinated proteins in wild-type and E3 ligase-deficient samples. For example, if a given protein were a substrate of Ube3A, then in principle it should be less ubiquitinated in samples specifically lacking this ubiquitin ligase. However, because ubiquitin is an extremely conserved molecule through evolution, it has been difficult to generate the high-affinity antibody reagents for this modification required for this proteomic approach.

To address this limitation, we employed a transgenic mouse strain in which a hemagglutinin epitope-tagged version of ubiquitin (HA-ubiquitin) is recombined into the endogenous *Hprt* locus, resulting in the widespread expression of this tagged ubiquitin at physiological levels (Ryu et al. 2007). Our studies indicate that HA-ubiquitin is effectively incorporated into substrates in these HA-ubiquitin transgenic animals, and immunoprecipitation with high-affinity anti-HA antibodies allows for the efficient recovery of ubiquitinated proteins from brain lysates prepared from these mice. To identify neuronal Ube3A substrates, these HA-ubiquitin transgenic mice were crossed with wild-type or *Ube3a* maternal-deficient animals (Fig. 2). Brain lysates were prepared from the resulting progeny and subjected to immunoprecipitation with anti-HA antibodies to isolate HA-ubiquitinated proteins. Quantitative mass spectrometry was then used to compare ubiquitinated proteins in wild-type and $Ube3a^{m-/p+}$ samples, searching for HA-ubiquitinated proteins whose abundance was decreased in $Ube3a^{m-/p+}$ samples.

Using this approach, we identified the protein Sacsin as a candidate Ube3A substrate. Peptides corresponding to ubiquitinated Sacsin were isolated in brain lysates of wild-type but not $Ube3a^{m-/p+}$ mice, suggesting that Sacsin might not be

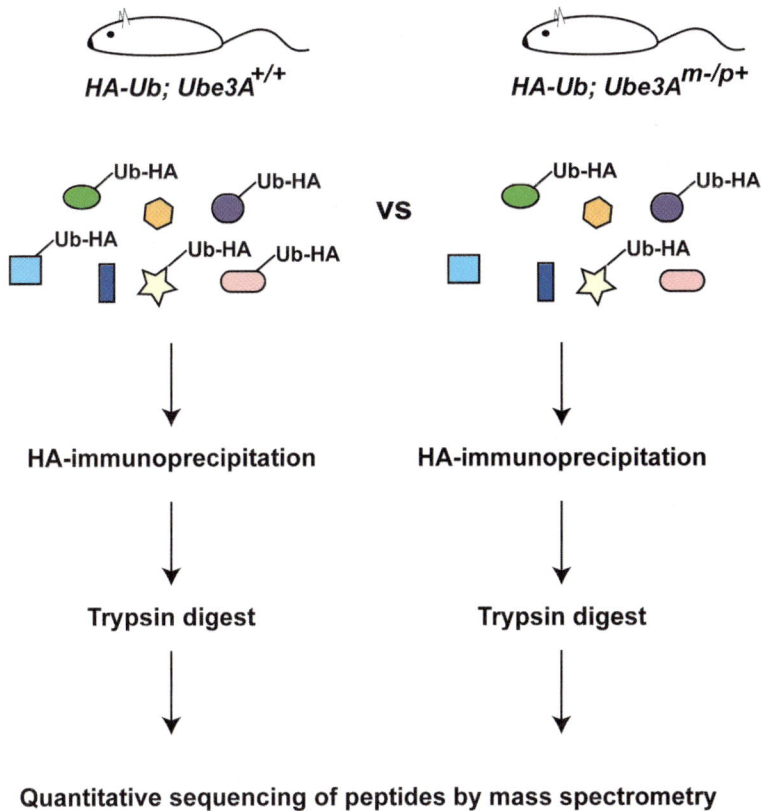

Fig. 2 A proteomic approach to Ube3A substrate identification. Brain lysates prepared from wild-type or *Ube3a* maternal-deficient transgenic HA-ubiquitin mice were subject to immunoprecipitation with anti-HA antibodies to isolate ubiquitinated proteins. Immunoprecipitates were digested with trypsin and the resulting tryptic peptides were subjected to quantitative mass spectrometry to identify proteins that exhibited reduced ubiquitination in the absence of Ube3A

efficiently ubiquitinated in the absence of Ube3A (Greer et al. 2010). While little is known regarding Sacsin's role in nervous system development, the *SACS* gene is mutated in the human disorder Charlevoix-Saguenay spastic ataxia, a neurological disorder with some phenotypic similarities to AS (Engert et al. 2000). Intriguingly, a region of Sacsin has been previously observed to share sequence homology with the XPC domain of the previously characterized Ube3A substrate HHR23A (Kamionka and Feigon 2004). As the specificity of ubiquitin ligases is most strongly determined by substrate binding, we hypothesized that this region of similarity between Sacsin and HHR23A might reflect a common mode of interaction with Ube3A. Consistent with this hypothesis, deletion of the putative HHR23A Ube3A-binding domain was found to significantly reduce both association of HHR23A with Ube3A and Ube3A-dependent HHR23A ubiquitination in vitro.

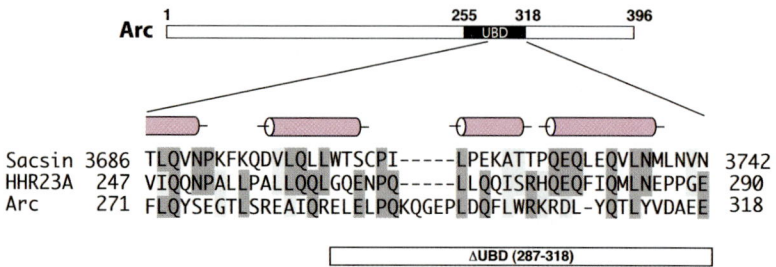

Fig. 3 Identification of a conserved Ube3A-binding domain. Sequence alignment of the Ube3A-binding domains of human HHR23A, Arc, and the highly related region present in the putative Ube3A substrate Sacsin. The location of conserved amphipathic helices identified in the HHR23A solution structure is shown (Kamionka and Feigon 2004), along with the extent of a deletion mutation (ΔUBD) sufficient to abrogate Arc binding to Ube3A. Note: some limited structural homology extends beyond the depicted area

Together, these results suggest the presence of a shared sequence motif on at least a subset of Ube3A substrates that mediates binding to Ube3A.

A bioinformatic search for mammalian proteins that possess sequence similarity to this motif identified a number of candidate Ube3A substrates, including Arc/Arg3.1 (activity-regulated cytoskeleton-associated; hereafter referred to as Arc), a synaptic protein that promotes the internalization of AMPA-type glutamate receptors (Fig. 3; Chowdhury et al. 2006; Rial Verde et al. 2006; Shepherd et al. 2006). Given our observation that Ube3A regulates AMPAR endocytosis at synapses, these findings raised the possibility that dysregulation of Arc in the absence of Ube3A mediates the effects of Ube3A loss on AMPAR trafficking. We therefore focused our efforts on the potential role of Arc as a novel neuronal Ube3A substrate.

5 Arc is a Neuronal Ube3A Substrate that Mediates the Effect of Ube3A on AMPAR Trafficking

Ube3A was found to effectively ubiquitinate Arc both in vitro using purified recombinant proteins and upon co-expression in a heterologous cell type. Co-immunoprecipitation experiments using mouse brain extracts also demonstrated that Arc and Ube3A interact in the intact brain (Greer et al. 2010). To determine if Ube3A ubiquitinates and triggers Arc degradation in neurons, we compared Arc expression in the brains of wild-type and $Ube3a^{m-/p+}$ mice. As the expression of both Ube3A and Arc is enhanced by neuronal activity, we exposed the mice to kainic acid or an enriched environment to boost the levels of Ube3A and Arc mRNA and protein. Under both of these conditions, we found that the level of Arc protein was significantly higher in $Ube3a^{m-/p+}$ mice than in wild-type controls (Fig. 4). In contrast to Arc, the activity-dependent phosphorylation of the transcriptional regulator MeCP2 and the induction of the activity-regulated transcription

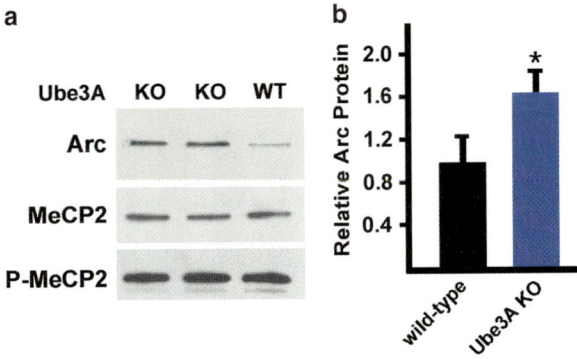

Fig. 4 Elevated Arc protein levels in $Ube3a^{m-/p+}$ mice. (**a**) Western blots of brain lysates from wild-type and $Ube3a^{m-/p+}$ mice that had been seized for 2.5 h by injection with kainic acid using anti-MeCP2, anti-phospho-MeCP2, and anti-Arc antibodies as indicated. Three independent experiments representing at least five animals per genotype were performed, and representative images are shown. (**b**) Quantification of Arc protein by Western blot analysis of brain lysates from wild-type and $Ube3a^{m-/p+}$ mice that had been exposed to an enriched environment for 2.5 h. Data represent mean ±SEM from four animals of each genotype. *Asterisk* denotes significance in pairwise comparison to control (P< 0.01, T-test). Adapted from Greer et al. (2010) with permission

factor NPAS4 remained unaffected. Furthermore, Arc mRNA levels were similar in the brains of wild-type and $Ube3a^{m-/p+}$ mice, indicating that the increase in the level of Arc protein detected in Ube3A-deficient neurons is likely due to a specific defect in Ube3A-mediated degradation rather than a change in the level of Arc mRNA. Given that Arc is ubiquitinated by Ube3A in vitro and in intact cells, and the level of Arc protein is significantly higher in $Ube3a$ maternal-deficient mice, we conclude that Arc is a bona fide Ube3A substrate and that the absence of Ube3A-dependent ubiquitination of Arc in $Ube3a^{m-/p+}$ mice results in increased levels of Arc in the brains of these animals.

As previously reported (Chowdhury et al. 2006; Rial Verde et al. 2006; Shepherd et al. 2006), overexpression of Arc was found to result in a decrease in cell surface GluR1 expression in cultured neurons, phenocopying the effect of Ube3A loss. Importantly, we found that the effect of Arc overexpression on GluR1 trafficking could be attenuated by coexpression of Ube3A. Although overexpression of a version of Arc lacking the putative Ube3A-binding domain still reduced cell surface GluR1 levels, in this case the effect was not reversed by coexpression of Ube3A. Thus, the ability of Ube3A to oppose Arc-driven GluR1 internalization appears to reflect Ube3A-mediated degradation of Arc.

To test whether Arc mediates the ability of Ube3A to regulate the number of AMPA receptors at synapses, hippocampal neurons were transfected with shRNAs targeting Ube3A to eliminate Ube3A and/or shRNA directed against Arc to reduce Arc expression (Fig. 5). Introduction of a shRNA specifically directed against Arc mRNA reversed the effect of Ube3A knockdown on surface GluR1 expression, suggesting that Ube3A promotes cell surface expression of AMPARs at synapses

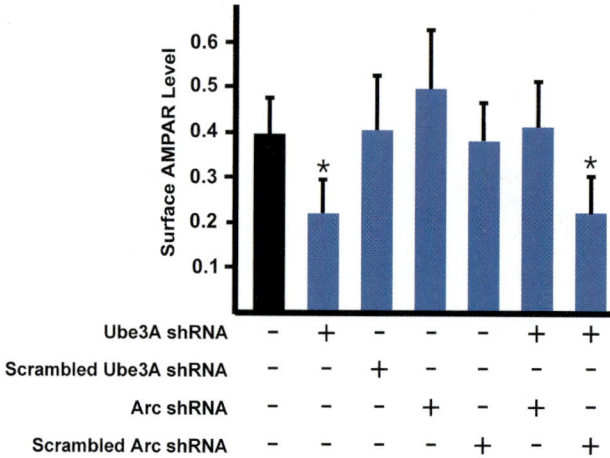

Fig. 5 Arc knockdown rescues the effects of Ube3A loss on synaptic development and function. Quantification of surface AMPAR expression in E18 + 14DIV hippocampal neurons transfected at 10 DIV with a GFP expression vector together with the indicated expression constructs. Data are presented as mean ±SEM from three independent experiments (*P < 0.01). Adapted from Greer et al. (2010) with permission

by ubiquitinating and degrading Arc and that, in the absence of Ube3A, there is an excess of Arc protein, resulting in an increased AMPAR internalization.

6 Discussion

While further experiments are still required, our current results are consistent with the following model. In the absence of robust synaptic stimulation, Arc and Ube3A are expressed at relatively low levels. In response to neuronal activity, Arc expression is induced with relatively rapid kinetics and functions to endocytose AMPARs from the plasma membrane. By contrast, Ube3A expression is induced with delayed kinetics and functions to ubiquitinate and degrade Arc, thus preventing excessive AMPAR internalization. In the absence of Ube3A, synaptic function is compromised by excessive AMPAR internalization due to abnormally high Arc levels (Fig. 6). These results suggest that Ube3A is an important component of the activity-dependent gene expression program that functions in concert with other activity-regulated genes to regulate synaptic development, which may account for the observation that Ube3A plays an important role in experience-driven synaptic plasticity. Moreover, these findings raise the possibility that elevated Arc levels contribute to at least some aspects of the cognitive dysfunction associated with AS.

AS is a debilitating neurodevelopmental disorder for which no effective treatment is currently available. The finding that the disruption of Ube3A activity results in a decrease in synaptic AMPAR expression suggests that interventions that

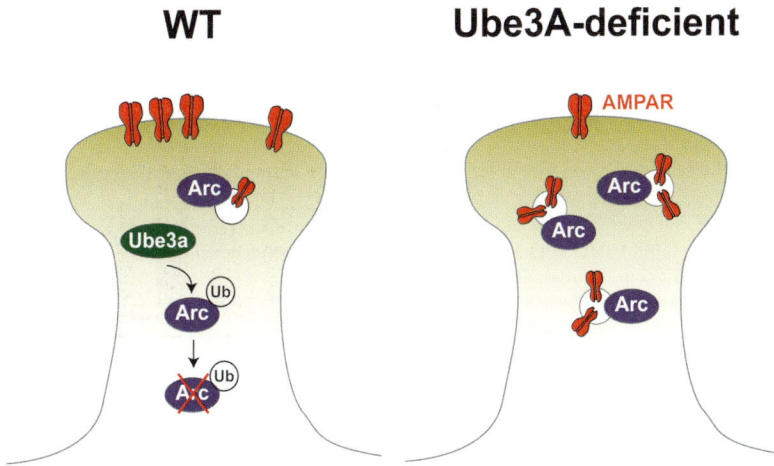

Fig. 6 Ube3A regulates AMPA receptor trafficking via the ubiquitination and degradation of Arc. During normal development (*left*), the cell surface levels of AMPA receptors are tightly controlled by Arc-mediated endocytosis. Arc expression is induced by neuronal activity, but activity-induced Ube3A acts to ubiquitinate and degrade the Arc protein. In AS-affected individuals (*right*), abnormally high levels of Arc protein accumulate due to the absence of neuronal Ube3A, resulting in synaptic dysfunction due to the excessive internalization of cell surface AMPA receptors. *WT* wild-type

promote cell surface AMPAR expression may represent potential therapeutics for treating this and related disorders. In this regard, Fragile X mental retardation syndrome (FXS) is associated with decreased cell surface AMPAR expression and elevated Arc levels, which can be partially reversed through the suppression of metabotropic glutamate receptor (mGluR) signaling (Bear et al. 2004; Zalfa et al. 2003). Indeed, genetic and pharmacological inhibition of mGluR signaling has recently been found to rescue a variety of FXS-associated phenotypes in mouse models (Dolen et al. 2007; Yan et al. 2005), and an mGluR5 antagonist (STX107) has been considered for human trials for the treatment of Fragile X syndrome. The finding that elevated Arc levels contribute to the synaptic phenotypes associated with the loss of Ube3A function in the central nervous system suggests that similar approaches might be therapeutic for AS as well.

Although these findings provide important insights into AS etiology, the extent to which elevated Arc expression contributes in vivo to the synaptic defects observed in AS-affected individuals remains to be determined. Indeed, given the broad phenotypic consequences of Ube3A mutations, it is likely that the dysregulation of a number of Ube3A substrates contributes to AS. For example, it is not obvious how impaired AMPAR function results in an increased susceptibility to seizures. The identification of a putative Ube3A-binding motif in substrate proteins should aid in the identification of additional Ube3A substrates. Future studies will seek to determine how these substrates work together with Ube3A to regulate both synaptic development and cognitive function.

References

Amir RE, Van den Veyver IB, Wan M, Tran CQ, Francke U, Zoghbi HY (1999) Rett syndrome is caused by mutations in X-linked MECP2, encoding methyl-CpG-binding protein 2. Nat Genet 23:185–188

Bear MF, Huber KM, Warren ST (2004) The mGluR theory of fragile X mental retardation. Trends Neurosci 27:370–377

Chowdhury S, Shepherd JD, Okuno H, Lyford G, Petralia RS, Plath N, Kuhl D, Huganir RL, Worley PF (2006) Arc/Arg3.1 interacts with the endocytic machinery to regulate AMPA receptor trafficking. Neuron 52:445–459

Cook EH, Lindgren V, Leventhal BL, Courchesne R, Lincoln A, Shulman C, Lord C, Courchesne E (1997) Autism or atypical autism in maternally but not paternally derived proximal 15q duplication. Am J Human Genet 60:928–934

Cooper EM, Hudson AW, Amos J, Wagstaff J, Howley PM (2004) Biochemical analysis of Angelman syndrome-associated mutations in the E3 ubiquitin ligase E6-associated protein. J Biol Chem 279:41208–41217

Dindot SV, Antalffy BA, Bhattacharjee MB, Beaudet AL (2008) The Angelman syndrome ubiquitin ligase localizes to the synapse and nucleus, and maternal deficiency results in abnormal dendritic spine morphology. Human Mol Genet 17:111–118

Dolen G, Osterweil E, Rao BS, Smith GB, Auerbach BD, Chattarji S, Bear MF (2007) Correction of fragile X syndrome in mice. Neuron 56:955–962

Engert JC, Berube P, Mercier J, Dore C, Lepage P, Ge B, Bouchard JP, Mathieu J, Melancon SB, Schalling M, Lander ES, Morgan K, Hudson TJ, Richter A (2000) ARSACS, a spastic ataxia common in northeastern Quebec, is caused by mutations in a new gene encoding an 11.5-kb ORF. Nat Genet 24:120–125

Flavell SW, Cowan CW, Kim TK, Greer PL, Lin Y, Paradis S, Griffith EC, Hu LS, Chen C, Greenberg ME (2006) Activity-dependent regulation of MEF2 transcription factors suppresses excitatory synapse number. Science 311:1008–1012

Flavell SW, Kim TK, Gray JM, Harmin DA, Hemberg M, Hong EJ, Markenscoff-Papadimitriou E, Bear DM, Greenberg ME (2008) Genome-wide analysis of MEF2 transcriptional program reveals synaptic target genes and neuronal activity-dependent polyadenylation site selection. Neuron 60:1022–1038

Glessner JT, Wang K, Cai G, Korvatska O, Kim CE, Wood S, Zhang H, Estes A, Brune CW, Bradfield JP, Imielinski M, Frackelton EC, Reichert J, Crawford EL, Munson J, Sleiman PM, Chiavacci R, Annaiah K, Thomas K, Hou C, Glaberson W, Flory J, Otieno F, Garris M, Soorya L, Klei L, Piven J, Meyer KJ, Anagnostou E, Sakurai T, Game RM, Rudd DS, Zurawiecki D, McDougle CJ, Davis LK, Miller J, Posey DJ, Michaels S, Kolevzon A, Silverman JM, Bernier R, Levy SE, Schultz RT, Dawson G, Owley T, McMahon WM, Wassink TH, Sweeney JA, Nurnberger JI, Coon H, Sutcliffe JS, Minshew NJ, Grant SF, Bucan M, Cook EH, Buxbaum JD, Devlin B, Schellenberg GD, Hakonarson H (2009) Autism genome-wide copy number variation reveals ubiquitin and neuronal genes. Nature 459:569–573

Greer PL, Greenberg ME (2008) From synapse to nucleus: calcium-dependent gene transcription in the control of synapse development and function. Neuron 59:846–860

Greer PL, Hanayama R, Bloodgood BL, Mardinly AR, Lipton DM, Flavell SW, Kim TK, Griffith EC, Waldon Z, Maehr R, Ploegh HL, Chowdhury S, Worley PF, Steen J, Greenberg ME (2010) The Angelman syndrome protein Ube3A regulates synapse development by ubiquitinating Arc. Cell 140:704–716

Greger IH, Esteban JA (2007) AMPA receptor biogenesis and trafficking. Curr Opin Neurobiol 17:289–297

Harris KM, Kater SB (1994) Dendritic spines: cellular specializations imparting both stability and flexibility to synaptic function. Annu Rev Neurosci 17:341–371

Jiang YH, Armstrong D, Albrecht U, Atkins CM, Noebels JL, Eichele G, Sweatt JD, Beaudet AL (1998) Mutation of the Angelman ubiquitin ligase in mice causes increased cytoplasmic p53 and deficits of contextual learning and long-term potentiation. Neuron 21:799–811

Kamionka M, Feigon J (2004) Structure of the XPC binding domain of hHR23A reveals hydrophobic patches for protein interaction. Protein Sci 13:2370–2377

Kishino T, Lalande M, Wagstaff J (1997) UBE3A/E6-AP mutations cause Angelman syndrome. Nat Genet 15:70–73

Kumar S, Talis AL, Howley PM (1999) Identification of HHR23A as a substrate for E6-associated protein-mediated ubiquitination. J Biol Chem 274:18785–18792

Malzac P, Webber H, Moncla A, Graham JM, Kukolich M, Williams C, Pagon RA, Ramsdell LA, Kishino T, Wagstaff J (1998) Mutation analysis of UBE3A in Angelman syndrome patients. Am J Human Genet 62:1353–1360

Matsuura T, Sutcliffe JS, Fang P, Galjaard RJ, Jiang YH, Benton CS, Rommens JM, Beaudet AL (1997) De novo truncating mutations in E6-AP ubiquitin-protein ligase gene (UBE3A) in Angelman syndrome. Nat Genet 15:74–77

Micheva KD, Smith SJ (2007) Array tomography: a new tool for imaging the molecular architecture and ultrastructure of neural circuits. Neuron 55:25–36

Miura K, Kishino T, Li E, Webber H, Dikkes P, Holmes GL, Wagstaff J (2002) Neurobehavioral and electroencephalographic abnormalities in Ube3a maternal-deficient mice. Neurobiol Dis 9:149–159

Morrow EM, Yoo SY, Flavell SW, Kim TK, Lin Y, Hill RS, Mukaddes NM, Balkhy S, Gascon G, Hashmi A, Al-Saad S, Ware J, Joseph RM, Greenblatt R, Gleason D, Ertelt JA, Apse KA, Bodell A, Partlow JN, Barry B, Yao H, Markianos K, Ferland RJ, Greenberg ME, Walsh CA (2008) Identifying autism loci and genes by tracing recent shared ancestry. Science 321:218–223

Patrick GN (2006) Synapse formation and plasticity: recent insights from the perspective of the ubiquitin proteasome system. Curr Opin Neurobiol 16:90–94

Petrij F, Giles RH, Dauwerse HG, Saris JJ, Hennekam RC, Masuno M, Tommerup N, van Ommen GJ, Goodman RH, Peters DJ, Breuning MH (1995) Rubinstein–Taybi syndrome caused by mutations in the transcriptional co-activator CBP. Nature 376:348–351

Rial Verde EM, Lee-Osbourne J, Worley PF, Malinow R, Cline HT (2006) Increased expression of the immediate-early gene arc/arg3.1 reduces AMPA receptor-mediated synaptic transmission. Neuron 52:461–474

Ryu KY, Maehr R, Gilchrist CA, Long MA, Bouley DM, Mueller B, Ploegh HL, Kopito RR (2007) The mouse polyubiquitin gene UbC is essential for fetal liver development, cell-cycle progression and stress tolerance. EMBO J 26:2693–2706

Sato M, Stryker MP (2010) Genomic imprinting of experience-dependent cortical plasticity by the ubiquitin ligase gene Ube3a. Proc Natl Acad Sci USA 107:5611–5616

Scheffner M, Huibregtse JM, Vierstra RD, Howley PM (1993) The HPV-16 E6 and E6-AP complex functions as a ubiquitin-protein ligase in the ubiquitination of p53. Cell 75:495–505

Shepherd JD, Rumbaugh G, Wu J, Chowdhury S, Plath N, Kuhl D, Huganir RL, Worley PF (2006) Arc/Arg3.1 mediates homeostatic synaptic scaling of AMPA receptors. Neuron 52:475–484

Sutcliffe JS, Nurmi EL, Lombroso PJ (2003) Genetics of childhood disorders: XLVII. Autism, part 6: duplication and inherited susceptibility of chromosome 15q11-q13 genes in autism. J Am Acad Child Adolesc Psychiatry 42:253–256

Trivier E, De Cesare D, Jacquot S, Pannetier S, Zackai E, Young I, Mandel JL, Sassone-Corsi P, Hanauer A (1996) Mutations in the kinase Rsk-2 associated with Coffin–Lowry syndrome. Nature 384:567–570

Veenstra-Vanderweele J, Christian SL, Cook EH (2004) Autism as a paradigmatic complex genetic disorder. Annu Rev Genomics Human Genet 5:379–405

Weeber EJ, Jiang YH, Elgersma Y, Varga AW, Carrasquillo Y, Brown SE, Christian JM, Mirnikjoo B, Silva A, Beaudet AL, Sweatt JD (2003) Derangements of hippocampal calcium/calmodulin-dependent protein kinase II in a mouse model for Angelman mental retardation syndrome. J Neurosci 23:2634–2644

Williams CA (2005) Neurological aspects of the Angelman syndrome. Brain Dev 27:88–94

Yan QJ, Rammal M, Tranfaglia M, Bauchwitz RP (2005) Suppression of two major Fragile X Syndrome mouse model phenotypes by the mGluR5 antagonist MPEP. Neuropharmacology 49:1053–1066

Yashiro K, Riday TT, Condon KH, Roberts AC, Bernardo DR, Prakash R, Weinberg RJ, Ehlers MD, Philpot BD (2009) Ube3a is required for experience-dependent maturation of the neocortex. Nat Neurosci 12:777–783

Zalfa F, Giorgi M, Primerano B, Moro A, Di Penta A, Reis S, Oostra B, Bagni C (2003) The fragile X syndrome protein FMRP associates with BC1 RNA and regulates the translation of specific mRNAs at synapses. Cell 112:317–327

Zoghbi HY (2003) Postnatal neurodevelopmental disorders: meeting at the synapse? Science 302:826–830

Targeting Children's Brain Tumors: Development of Hedgehog Pathway Inhibitors for Medulloblastoma

Tom Curran

Abstract Hedgehog (Hh) is a secreted protein family that controls proliferation, differentiation, cell fate specification, left-right asymmetry and morphogenesis during development. The Hh pathway is extremely complex and mutations in various components result in developmental abnormalities as well as increased tumor incidence, including medulloblastoma, in mice and humans. Loss of one copy of Patched 1 (Ptch1), a receptor for Hh, causes increased activity of Smoothened (Smo) and elevated expression of the transcription factor Gli1 in both familial and sporadic medulloblastoma. Fifteen years ago, we embarked on a long-term program to develop new therapeutic approaches for this devastating childhood brain tumor. Small molecule inhibitors of the Hh pathway, both naturally occurring and chemically synthesized, inhibit Smo activity and provide potential novel therapeutic agents. We generated a genetically engineered mouse medulloblastoma model, $Ptch1^{+/-}p53^{-/-}$ mice, that exhibits a 100% incidence of medulloblastoma. Using these mice, we tested the effect of a Smo inhibitor (HhAntag), isolated by Curis Inc., on the growth of spontaneous medulloblastoma. We found that twice daily dosing, by oral gavage, effectively eliminated even large tumor masses in the cerebellum of mice 3 to 10 weeks of age. These findings led to Phase I clinical trials of a Smo inhibitor in pediatric and adult medulloblastoma.

1 Introduction

Pediatric brain tumors are quite distinct from adult brain tumors and they appear to be derived from progenitor cell populations that exist only transiently in the developing brain. However, with a few exceptions, it has been difficult to delineate the normal progenitor cells from which most of these tumors arise, which implies

T. Curran
Department of Pathology and Laboratory Medicine, The Children's Hospital of Philadelphia, Philadelphia, PA 19104, USA
e-mail: currant@email.chop.edu

T. Curran and Y. Christen (eds.), *Two Faces of Evil: Cancer and Neurodegeneration*, Research and Perspectives in Alzheimer's Disease, DOI 10.1007/978-3-642-16602-0_5, © Springer-Verlag Berlin Heidelberg 2011

that the signaling pathways responsible for driving tumor growth may have their origins in neurodevelopment. Thus, the molecular targets and pathways critical for brain tumors in children may only partially overlap with those identified in adult disease. Importantly, the context of tumor growth is quite distinct; therefore, it is critical to consider the developmental environment while investigating the biology of childhood brain tumors and, particularly, while testing potential therapeutic intervention strategies. Medulloblastoma is a somewhat inappropriate name for the most common malignant brain tumor of childhood. The peak age at presentation is 7 years, but tumors arise throughout childhood, with 70% occurring in children younger than 16 (Louis et al. 2007). Once considered related to glioma, medulloblastoma were proposed to arise from undifferentiated embryonal cells, or "medulloblasts," located in the ependymal lining of the fourth ventricle (Bailey and Cushing 1925). Although this is no longer thought to be the case, the introduction of histopathological classification of tumors by Bailey heralded a new age in neuro-oncology, and the recognition of medulloblastoma as a distinct tumor type had a major impact on disease management. In a sense, the current molecular characterization of brain tumors, by genotyping and DNB sequence analyses, represents a further refinement of these early principles of disease classification. The ultimate goal of this approach is to define specific categories of disease, based on their molecular signatures, for stratification and treatment with agents directed against distinct targets appropriate for each subset.

Historically, pediatric brain tumors have received little attention from the pharmaceutical industry because of their limited market potential. Although patient survival rates have benefited from advances made in radiation delivery as well as in surgical technique, and the use of certain combinations of chemotherapeutics in the adjuvant setting has reduced recurrence, the picture remains dismal because of the long-term sequelae of current therapy. Standard treatment, which involves a combination of surgery, radiation and chemotherapy, is relatively successful, with a 5-year survival rate of up to 78%. However, the prognosis is much worse for patients younger than three and in older patients with metastatic disease. There is also significant morbidity associated with treatment, including cognitive, endocrine and neuropsychological deficits, particularly in younger patients. Thus, there is a great need for alternative therapies. We pursued a long-term strategy designed to bring new agents to the clinic. The goal was to utilize genomic approaches to further define the molecular subtypes of medulloblastoma and uncover specific targets (Pomeroy et al. 2002; Thompson et al. 2006). We hoped to develop mouse models appropriate for disease subsets and to use these to test novel therapeutics prior to introduction in a clinical trial. A number of advances in the field led to identification of the Hedgehog (Hh) pathway as a target in 30% of medulloblastoma, which allowed us to develop an appropriate mouse model to test novel inhibitors. The success of this approach in mice supported the introduction of a novel therapeutic in a Phase I clinical trial in pediatric medulloblastoma that concluded in 2010. The following report outlines some of the key steps in this process of shepherding a novel concept from the basic research laboratory through animal models and into the clinic. This journey was inspired by interactions with several young patients who, despite their unfortunate circumstances, remained remarkably upbeat and positive about the future.

2 Role of the Hh Pathway in Medulloblastoma

The discovery that the receptor of Hh signaling, Ptch1, is mutated in a subset of spontaneous and basal cell nevus syndrome (BCNS)-associated medulloblastoma provided the first major link to the etiology of this disease (Gorlin 1987; Hahn et al. 1996; Johnson et al. 1996; Pietsch et al. 1997; Raffel et al. 1997; Xie et al. 1997). BCNS, also known as Gorlin Syndrome; OMIM 109400, is an autosomal dominant disease (Gorlin 1987) characterized by larger body size, developmental and skeletal anomalies, fibromas of soft tissues, radiation sensitivity, basal cell carcinoma, and medulloblastoma (Goodrich and Scott 1998; Kimonis et al. 1997). Ptch1 encodes a 12-pass transmembrane protein that functions as a receptor complex for Hh (Murone et al. 1999). Ptch1 represses constitutive signaling by Smo, which activates transcription of the zinc finger protein Gli1 (Fig. 1). The Shh pathway in vertebrates is complicated by the existence of at least one Ptch1-related gene,

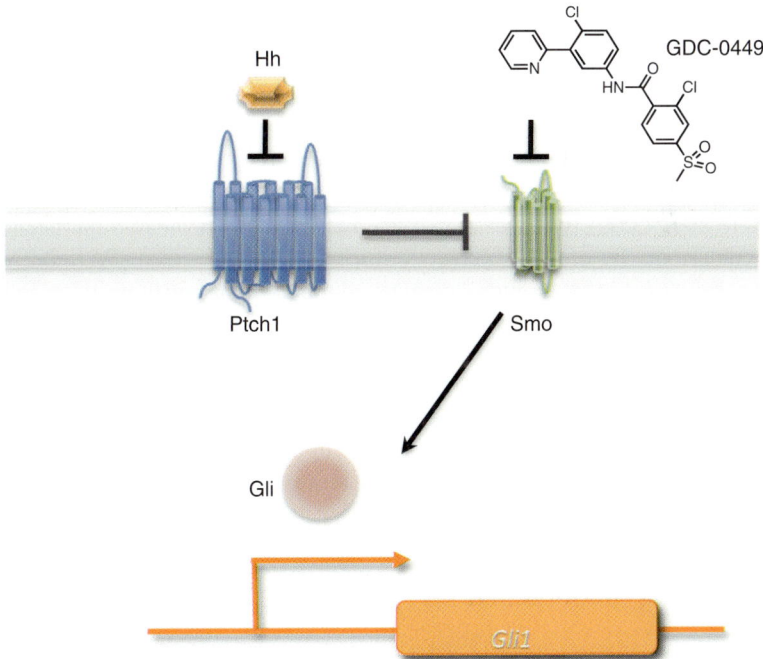

Fig. 1 Diagram of the Hh pathway. Hh ligands bind to Ptch1 and inhibit its ability to block Smo. GDC-0449 and the other inhibitors of the Hh pathway currently being tested in the clinic bind and inactivate Smo. Smo signals through Gli proteins to regulate transcription of target genes, including Gli1

Ptch2, three Hh-related genes (referred to as Sonic (Shh), Indian (Ihh) and Desert (Dhh) hedgehog), and three Gli genes (Gli1, Gli2 and Gli3). There are also many other genes that influence the modification, transport and processing of Hh and Gli proteins, all of which represent potential targets for intervention (Rubin and de Sauvage 2006). The mechanism responsible for regulation of Smo by PtchI varies a little between insect and mammalian cells; however, recent results have shed light on the processes involved. In the absence of Hh, Ptch1 is localized to cilia (Huangfu et al. 2003) where it has been proposed to exclude Smo, allowing cilia to function as chemosensors for Shh (Rohatgi et al. 2007). In this model, Hh binding transports Ptch1 out of the cilia, permitting entry of Smo, which then undergoes activation. Rohatgi et al. (2007) further suggested a candidate class of molecules for the proposed natural regulators of Smo (Taipale et al. 2002). They demonstrated that certain oxysterols promoted translocation of Smo into cilia regardless of the presence of Ptch1. However, the oxysterols appeared to accomplish this translocation without binding to Smo.

Activating mutations in Smo have been observed in sporadic medulloblastoma, and alterations in suppressor-of-fused (Sufu) have been reported both in sporadic disease and in families predisposed to medulloblastoma (Taylor et al. 2002). In mice, loss of Sufu promotes medulloblastoma formation (Lee et al. 2007). Inactivation of Ptch1 abrogates its repressor function thereby increasing Gli1 expression, which was first identified as an amplified gene in glioblastoma (Kinzler et al. 1987). Gli1 is not readily detected in adult brain RNA but it is highly expressed in granule neuron progenitor cells. Over-expression of Gli1 in fibroblasts causes cell transformation, and the expression of Gli1 in skin results in lesions resembling human basal cell carcinoma (BCC; Goodrich and Scott 1998).

In mice, a heterozygous mutation of Ptch1 is associated with an approximately 20% incidence of medulloblastoma over a period of 1 year (Goodrich et al. 1997; Wetmore et al. 2000). We accelerated the incidence and decreased the latency of medulloblastoma by crossing $Ptch1^{+/-}$ mice with $p53^{-/-}$ mice (Wetmore et al. 2001). All $Ptch1^{+/-}p53^{-/-}$ mice develop medulloblastoma and most die from brain tumors within 12 weeks of birth. P53 is mutated in approximately 10% of human medulloblastoma, and the p53 pathway is defective in a larger subset of tumors (Thompson et al. 2006). In addition, germline mutation of P53 is associated with medulloblastoma in Li–Fraumeni syndrome (Li and Fraumeni 1969). Thus, $Ptch1^{+/-}p53^{-/-}$ mice provide an ideal model for testing new treatments for medulloblastoma caused by activation of the Hh pathway.

3 Targeting Smoothened

Potential therapeutic agents, specific for the Hh pathway, were identified initially as the alkaloid teratogens present in corn lilies, cyclopamine and jervine, that cause holoprosencephaly when ingested by pregnant sheep during a critical period in the first trimester (Binns et al. 1963, 1965; Keeler, 1972). Subsequently, they were shown to function by binding to Smo to block downstream signaling, resulting in inhibition of proliferation (Chen et al. 2002; Cooper et al. 1998). Several structurally unrelated compounds have now been isolated from small molecule screens using cell-based assays, which function by binding to Smo competitively with cyclopamine (Chen et al. 2002; Frank-Kamenetsky et al. 2002; Gabay et al. 2003; Williams et al. 2003). In addition, agonist compounds have been identified that bind to the same site on Smo but result in increased activity in the absence of Hh (Chen et al. 2002; Frank-Kamenetsky et al. 2002).

There are several potential mechanisms whereby the Hh pathway activity can promote tumorigenesis: *Type I;* constitutive activation of the Hh pathway, resulting in increased expression of target genes, including Gli1 and Gli2. This can occur by loss of function of Ptch1 (deletion or suppression), gain of function of Smo, or loss of function of SUFU; *Type II*: autocrine stimulation by Hh ligands produced in tumor cells; *Type III*: paracrine stimulation of tumor cells by Hh ligands secreted from surrounding stromal cells; and *Type IV*: paracrine support of the tumor niche (stromal cells and vasculature) by Hh ligands secreted from tumor cells causing release of factors such as IGF and VEGF that promote tumor growth. To date, the only clear classes of tumors that fit the *Type I* category are basal cell carcinoma (BCC) and medulloblastoma, and these are the only two tumor types that show dramatic responses to treatment in authentic preclinical models and in patients. Many tumors have been suggested to be in the *Type II* category; however, in most cases it is difficult to distinguish among *Types II, III,* and *IV*, as it would require careful analysis of the expression patterns of Hh pathway components and target genes independently in tumor cells and stromal cells (Yauch et al. 2008). Nevertheless, the expression of Hh ligands in many tumor populations, including gastrointestinal, liver, breast, lung, prostate, pancreatic and glioma cancers, has led to proposals that these may be effectively treated using Hh pathway antagonists (Rubin and de Sauvage 2006). This broad potential market for Smo antagonists has attracted a great deal of attention from major pharmaceutical companies, as well as many smaller biotech companies. Many now include the Hh pathway in their portfolio of oncology targets. Several academic laboratories utilized the naturally occurring inhibitor, cyclopamine, to target the Hh pathway. However, there are many shortcomings with this compound, particularly a high level of non-specific toxicity, which confound its use in preclinical animal studies. Therefore, in 2001, I contacted a small biotech company, Curis Inc., with a proposal to investigate the use of their Hh pathway inhibitors in medulloblastoma. They agreed to a collaboration that allowed access to novel inhibitors and we launched a long-term interaction.

4   Hh Pathway is Downregulated in Tumor Cell Lines

One of the first surprises that impacted the direction of the project, was the observation that Hh pathway activity is suppressed in medulloblastoma cells cultured in vitro (Romer et al. 2004; Sasai et al. 2006). We made more than 20 independent tumor lines and invariably found that they all switched off expression of Hh pathway target genes (such as Gli1) very rapidly when cultured as monolayer cells. We also found they were no longer capable of responding to Shh or to small molecule agonists of Smo, which precluded the use of Smo antagonists in vitro to investigate the role of Hh pathway signaling in tumor cell growth and focused our efforts on the in vivo model system. The results were surprising as they were in stark contrast to the claims suggesting that cyclopamine (a Smo antagonist) specifically blocks growth of mouse medulloblastoma cell lines in vitro and following transplantation into flank allografts (Berman et al. 2002). Subsequently, we demonstrated that the change in tumor cells is profound and appears to be a consequence of a major epigenetic switch that occurs during the explantation process (Sasai et al. 2006). Furthermore, Hh pathway activity could not be restored by transplantation of tumor cell lines into the flank of immunosuppressed mice. We found that cyclopamine is a highly toxic compound, in vitro and in vivo, and that the concentrations used to suppress Hh signaling by other laboratories (between 1 and 10 μM) result in significant non-specific inhibition of cell proliferation. At least 1 μM cyclopamine is required to completely suppress Shh-stimulated activity in NIH3T3 cells, so the therapeutic index of cyclopamine is very narrow. In contrast, HhAntag completely blocks Hh signaling at a concentration of 100 nM and it only exhibits non-specific growth-inhibitory effects when used at levels above 10 μM (Romer et al. 2004). In vivo, we found that the toxic side effects of systemic cyclopamine, resulting in dehydration and weight loss, preclude it from being used at the dose required to completely suppress Hh pathway activity in young mice. Attempts to avoid the systemic toxicity of cyclopamine by injecting it directly into small flank tumors are prone to artifact (Curran 2010).

5   Direct Allografts Retain Hh Pathway Activity

It is frequently stated that tumor cell lines, particularly brain tumor cell lines, do not represent the in vivo tumor cell phenotype and, therefore, cannot be used to predict drug responses accurately. In the case of mouse medulloblastoma, we demonstrated specifically that this is a consequence of suppression of the Hh signaling pathway. In contrast, allografts derived directly from minced tumor tissue, which was never cultured in a dish, retain pathway activity and respond to treatment with HhAntag (Sasai et al. 2006). Using mouse medulloblastoma, we compared paired samples in gene expression microarray experiments, matching the original tumor, cell lines derived from the tumor, allografts derived from cell lines, and allografts derived from fresh tumors by direct transplant (Sasai et al. 2006). We found that the direct allografts were very similar to the original tumors in terms of the overall patterns of gene expression; however, the cell lines and allografts were extremely different, and this difference was not restricted to Hh pathway target genes. Furthermore, we found that allografts also differed dramatically from the cell lines from which they were made. These findings demonstrate the dangers inherent in the use of cell culture, and of tumor cell transplantation models derived from cultured cells, for preclinical analyses of potential therapies. It is likely that human tumor cell lines suffer from the same drawback. In fact, we found that the Hh pathway is inactive in all of the human medulloblastoma lines we examined. This finding has precluded any meaningful studies on the effects of Hh pathway antagonists in human medulloblastoma cell lines and xenografts.

6   Testing Targeted Therapies in Mice

Due to the absence of good cell line or xenograft models for medulloblastoma, we had to utilize the genetic mouse model for proof-of-concept studies of Hh pathway inhibitors. We developed a three-step system for testing targeted therapies in spontaneous mouse models. $Ptch1^{+/-}p53^{-/-}$ mice were chosen for this analysis because they are a highly penetrant, early onset, medulloblastoma model with 100% disease incidence within 2 weeks. This predictive model allowed the use of age-matched cohorts of mice. The first step was to demonstrate that the drug (HhAntag) passes the blood-brain barrier and inhibits the pathway, causing dose-dependent suppression of Gli1 as well as other Hh pathway target genes in tumor cells. We demonstrated that the pathway could be completely suppressed in tumors after giving HhAntag twice daily at 100 mg/kg by oral gavage for a period of 4 days (Romer et al. 2004). The second step was to demonstrate that pathway inhibition results in a tumor response. We found that a few days of treatment resulted in complete cessation of tumor cell proliferation, increased expression of neuronal differentiation markers and elevated levels of cell death. After 2 weeks of treatment, mice given HhAntag twice daily at 100 mg/kg, starting at 3 weeks of age, exhibited no residual tumor mass (Romer at al. 2004). There was some evidence of glial scarring, indicating that tumor mass had been present but was eradicated by the treatment. In the final step, we demonstrated that treatment significantly extended the survival of mice (Romer at al. 2004).

7   HhAntag Eradicates Large Tumors in $Ptch1^{+/-}p53^{-/-}$ Mice

The majority of $Ptch1^{+/-}p53^{-/-}$ mice die before the age of 12 weeks as a result of medulloblastoma. In our published studies, we demonstrated that tumors could be eliminated from relatively young mice, up to 6 weeks of age, after a 2-week course of therapy. However, we were also interested in determining whether HhAntag can eradicate even large tumor masses in the brain. Therefore, we treated 9 to 10 week-old $Ptch1^{+/-}p53^{-/-}$ mice with HhAntag for either 8 days or 2 weeks with 100 mg/kg HhAntag. The results demonstrated that HhAntag eradicated extremely large tumor volumes in a brief course of treatment. At this age, the tumors completely cover the cerebellum in the majority of mice examined (Fig. 2).

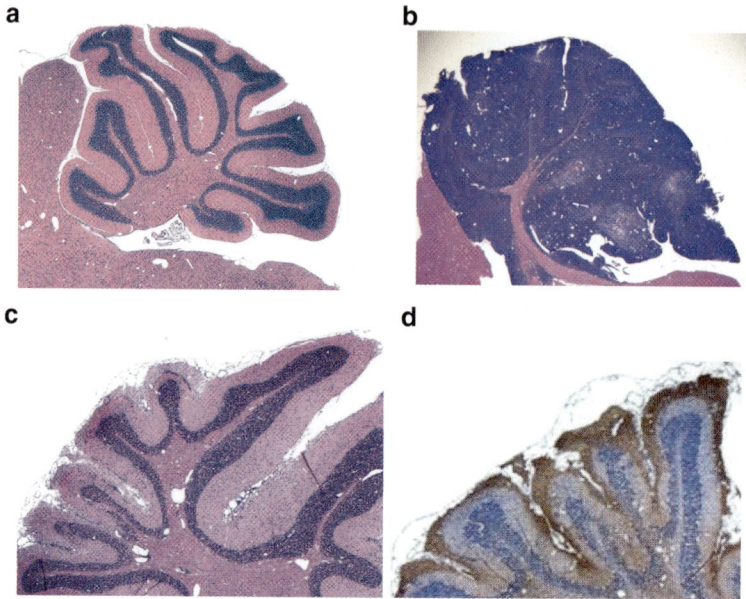

Fig. 2 HhAntag eradicates large brain tumors. Ten-week-old $Ptch1^{+/-}p53^{-/-}$ mice were treated for 2 weeks with 100 mg/kg HhAntag twice daily by oral gavage. Mice were perfused with paraformaldeyde and sections were prepared. H&E stains of normal cerebellum (**a**), untreated cerebellum with medulloblastoma (**b**) and cerebellum 2 weeks after treatment with HhAntag (**c**) are shown. Evidence of glial scarring can be seen by staining with GFAP (**d**)

Although the large tumors invade the cerebellum, a relatively intact cerebellar structure was revealed after treatment with HhAntag. Staining with GFAP revealed extensive gial scarring after treatment, indicating that substantial tumor mass had been present (Fig. 2), but there were no obvious behavioral or locomotor problems in the treated mice.

The result described above encouraged us to develop an indicator mouse based on Gli transcriptional activity that could be utilized in preclinical functional imaging studies of Hh pathway antagonists. Hh activity can be readily monitored by optical imaging in cells transfected with a plasmid containing eight Gli binding sites fused to the d-crystallin basal promoter and the firefly luciferase gene (Kamachi and Kondoh 1993; Sasaki et al. 1997). We used this same construct to generate six independent lines of transgenic mice by microinjection (Kimura et al. 2008). Primary granule cell cultures were prepared from litters obtained from each strain and tested for their ability to respond to Shh. Five lines did not show any change in luciferase activity in response to Shh, perhaps because transgene expression was suppressed by sequences surrounding the site of integration. However, one line exhibited a robust Shh-induced increase in luciferase activity, and this line was selected to generate a cohort of GliLuc mice for future studies. The Hh pathway is very active during embryogenesis and postnatal central nervous system development (McMahon et al. 2003). As expected, GliLuc transgenic embryos (visualized through the skin of the dam) exhibited a robust luciferase signal within 15 min of inoculating the dam with 150 mg/kg D-Luciferin by intraperitoneal (IP) injection (Kimura et al. 2008). As gestation progressed, GliLuc transgenic embryos showed increased luciferase activity. Embryos isolated at E17.5 were positive, with a particularly strong signal emanating from the developing brain. To determine the ability of the transgene to respond to Shh stimulation, mouse embryo fibroblast (MEF) cultures were prepared and treated with Shh for 48 h before measuring the level of luciferase activity. As little as 50 ng/ml Shh caused a dramatic increase in luciferase activity when measured by bioluminescence imaging or chemilumines-cence biochemical assays (Kimura et al. 2008). The maximal response was obtained with 100 ng/ml Shh, which is identical to the concentration required to promote maximal proliferation of primary granule cells in culture. HhAntag effec-tively blocked Hh signaling activated by 100 ng/ml Shh (Romer et al. 2004). The ability of HhAntag to suppress Hh pathway activity in embryos was investigated by treating pregnant dams with HhAntag at 100 mg/kg twice daily by oral gavage at both E13.5 and E14.5. The signals observed from treated embryos were diminished compared to those from untreated embryos. At the end of the study we examined the embryos and found evidence of shortened limbs as a consequence of exposure to HhAntag. To examine Hh pathway activity during postnatal development, we quantified the GliLuc signal in pups from postnatal days 1, 7 and 14, as well as in young adults at 5 and 11 weeks of age. Hh pathway activity was very high shortly after birth, and then it progressively declined throughout the course of postnatal development. Little or no GliLuc signal was detected in 11-week-old mice. Although the imaging approach mostly detects superficial signals in skin tissue, the skin still exhibited the highest level of pathway activity when compared with other dissected mouse tissues including, brain, liver, lung, kidney, heart, and spleen. During early mouse postnatal development, epidermal and dermal progenitors actively proliferate to form hair and non-hair epithelium as well as stratified epidermis (Fuchs 2007), and proliferation of these progenitors cells is under tight control by the Hh pathway (Callahan and Oro 2001).

8 Hh Pathway Inhibition Causes Bone Defects in Young Mice

In our previous studies in adult mice, we did not observe any weight loss or growth retardation following twice daily treatment with 100 mg/kg HhAntag for 2 weeks. We also dosed some mice once a day for more than 100 days with 100 mg/kg HhAntag with no deleterious effects. Therefore, we were surprised to note that young mice treated with HhAntag from P10 to P14 did not thrive. In fact, they lost weight during the period of drug treatment (day 10–14). Although the HhAntag-treated mice regained weight after cessation of treatment, they never caught up with their littermates, and they remained noticeably smaller than control mice even long after the brief 4-day treatment period, suggesting that short-term treatment of young mice with HhAntag permanently altered growth.

X-ray analysis revealed widespread abnormalities of the skeleton in GliLuc mice treated with HhAntag. In all treated mice, we found shortened endochondral bones and several displayed grossly abnormal bone structures, particularly the femur and tibia. Whole skeleton staining with Alizarin Red S revealed significant joint abnormalities similar to those reported in young mice in which Ihh is conditionally mutated at postnatal day 0 (Maeda et al. 2007). Imaging and histological analysis (Fig. 3) of sections of limbs from 12-week-old mice previously treated with HhAntag between P10–P14 revealed malformation of the epiphysis and growth plate. The columnar organization of chondrocytes in the growth plate was disrupted, and the cartilage structure appeared dysplastic. The extensive defects in bone growth were not accompanied by any other major defects in other organs, as determined by gross examination and necropsy of 12-week-old treated mice. These observations are consistent with the critical role the Hh pathway is known to play in

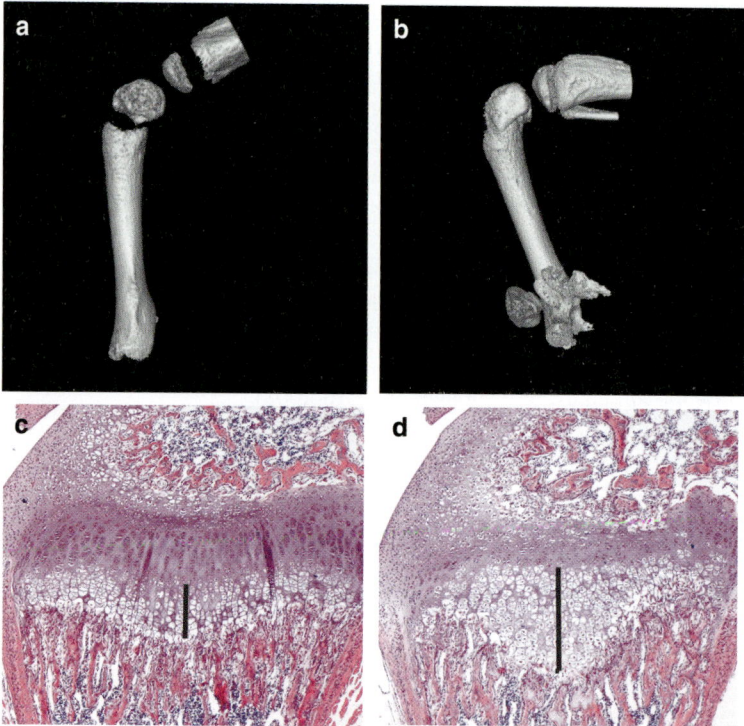

Fig. 3 HhAntag treatment of young mice results in bone defects. MicroCT imaging of the developing femur shows that treatment of 10-day-old mice for 4 days with 100 mg/kg HhAntag results in fusion of the growth plate. Untreated (**a**), treated (**b**); *arrows* indicate the position of the growth plate. H&E staining of the growth plate from untreated (**c**) and treated (**d**) mice reveals that the hypertrophic zone (*black bars*) is expanded in the absence of Hh signaling

bone development (Ehlen et al. 2006). Deletion of Ihh causes embryonic lethality in mice (St-Jacques et al. 1999), conditional ablation results in a phenotype similar to that reported here (Maeda et al. 2007), and hypomorphic mutations of *IHH* in humans cause acrocapitofemoral dysplasia (Hellemans et al. 2003). Next, we investigated the effects of HhAntag treatment duration on bone development in normal mice. Mice were treated with 100 mg/kg HhAntag at P10 twice daily for 1–4 days and the effects on bone development were analyzed at 6 weeks of age. We observed an obvious decrease in the growth of the long bones after only 2–4 days (four to eight doses of HhAntag) of treatment. More modest effects were observed on bone morphology in mice receiving only two doses of HhAntag over a 24-h period. The results raise concerns about the use of HhAntag and other Hh pathway inhibitors in pediatric patients. Elimination of tumors requires twice daily treatment of mice for approximately 2 weeks, whereas significant problems in bone development were observed after as little as 2 days.

9 Inhibiting the Hh Pathway in the Clinic

The success of Hh pathway inhibitors in the preclinical studies of medulloblastoma described above, and in cell culture models of BCC, encouraged Curis Inc. to establish a partnership with Genentech Inc. in 2005 to develop these novel agents as clinical anti-cancer drugs. The first Hh antagonist to be used in a clinical trial was the Curis compound aminoproline, Cur-61414, which was tested as a topical treatment for BCC. This compound was very effective in mouse models as a topical agent and in human BCC explants in culture (Williams et al. 2003). Unfortunately, the Phase I clinical trial was halted prematurely based on preliminary data showing no clinical activity. The reason for the failure was that, in contrast to mouse skin, Cur-61414 failed to penetrate human skin and downregulate Hh pathway target genes in BCC lesions. However, no toxicity problems were encountered. Although disappointing, this result did not hinder a medicinal chemistry program at Genentench that led to the lead compound, GDC-0449 (Robarge et al. 2009).

GDC-0449 showed remarkable responses in a Phase I clinical trial of locally advanced and metastatic solid tumors. In particular, patients with BCC tumors showed positive responses, leading to the recruitment of more BCC patients and expansion of the trial. Thirty-three patients with either locally advanced BCC or metastatic disease were treated with varying doses of GDC-0449 (Von Hoff et al. 2009). The response rate for patients with locally advanced BCC was 60% and the response rate for metastatic patients was 50%. Furthermore, 90% of these patients responded within four months after the initiation of treatment (55% within two months). Because of this unprecedentedly high response rate, a Phase II trial was initiated. With the exception of a single patient with medulloblastoma, no other tumor responses were described. The medulloblastoma case involved an adult patient with advanced metastatic disease (Rudin et al. 2009). Just 2 months after treatment initiation, PET scans showed a dramatic reduction in disease.

However, the patient relapsed because the tumor acquired a mutation in SMO that prevented binding of GDC-0449 to Smo (Yauch et al. 2009). This example was both a strong proof of concept of Smo as a target in medulloblastoma and a warning that resistant variants can arise quite readily. Nevertheless, both pediatric and adult MB Phase I trials were launched and positive results were reported without any major side effects (Gajjar et al. 2010).

The rationale for the use of Hh pathway inhibitors in BCC and medulloblastoma is very strong, since these tumors harbor activating mutations in the pathway. However, a series of studies in cell culture and xenograft mouse models also made a case that the same inhibitors would work in a broad range of tumors that lacked mutations in the pathway. In fact, some investigators predicted that the Hh pathway was involved in 25% of all human cancer (Lum and Beachy 2004), which led several companies, including Infinity, Bristol Myers Squibb, Novartis and Pfizer, to launch clinical trials of novel Hh pathway inhibitors (reviewed in Mas and Altaba 2010). Analysis of the experimental protocols used in some of these preclinical studies of other tumors revealed that they employed flawed approaches that potentially gave false positive results, for example, in colon carcinoma models (Curran 2010). Indeed, recently a Genentech trial of GDC-0449 in colon carcinoma was closed prematurely due to lack of response. Thus, while there is great hope that inhibition of the Hh pathway will contribute to cancer therapy in the future, this may only be the case in subsets of tumors. The lesson from medulloblastoma indicates that it is important to understand the specific role of the Hh pathway in molecular subtypes of tumor. In addition, it is critical to develop an animal model that reflects the biology of the corresponding human disease, particularly in the case of brain tumors. Hopefully, by understanding the many growth control pathways discussed in this provocative and exciting symposium, which govern the proliferation and survival of cells in the brain, we will also uncover additional therapeutic approaches to target additional specific subsets of brain tumors that cause such terrible devastation in patients.

References

Bailey P, Cushing H (1925) Medulloblastoma cerebelli. A common type of midcerebellar glioma of childhood. Arch Neurol Psychiat (Chic) 14:192–224
Berman DM, Karhadkar SS, Hallahan AR, Pritchard JI, Eberhart CG, Watkins DN, Chen JK, Cooper MK, Taipale J, Olson JM, Beachy PA (2002) Medulloblastoma growth inhibition by hedgehog pathway blockade. Science 297:1559–1561
Binns W, James LF, Shupe JL, Everett G (1963) A congenital cyclopian-type malformation in lambs induced by maternal ingestion of a range plant, Veratrum californicum. Am J Vet Res 24:1164–1175
Binns W, Shupe JL, Keeler RF, James LF (1965) Chronologic evaluation of teratogenicity in sheep fed Veratrum californicum. J Am Vet Med Assoc 147:839–842
Callahan CA, Oro AE (2001) Monstrous attempts at adnexogenesis: regulating hair follicle progenitors through Sonic hedgehog signaling. Curr Opin Genet Dev 11:541–546

Targeting Children's Brain Tumors: Development of Hedgehog Pathway Inhibitors   69

Chen JK, Taipale J, Cooper MK, Beachy PA (2002) Inhibition of Hedgehog signaling by direct binding of cyclopamine to Smoothened. Genes Dev 16:2743–2748

Cooper MK, Porter JA, Young KE, Beachy PA (1998) Teratogen-mediated inhibition of target tissue response to Shh signaling. Science 280:1603–1607

Curran T (2010) Mouse models and mouse supermodels. EMBO Mol Med 2010 Aug 18 DOI: 10.1002/emmm.201000090

Ehlen HW, Buelens LA, Vortkamp A (2006) Hedgehog signaling in skeletal development. Birth Defects Res C Embryo Today 78:267–279

Frank-Kamenetsky M, Zhang XM, Bottega S, Guicherit O, Wichterle H, Dudek H, Bumcrot D, Wang FY, Jones S, Shulok J, Rubin LL, Porter JA (2002) Small-molecule modulators of Hedgehog signaling: identification and characterization of Smoothened agonists and antagonists. J Biol 1:10

Fuchs E (2007) Scratching the surface of skin development. Nature 445:834–842

Gabay L, Lowell S, Rubin LL, Anderson DJ (2003) Deregulation of dorsoventral patterning by FGF confers trilineage differentiation capacity on CNS stem cells in vitro. Neuron 40:485–499

Gajjar AJ, Stewart CF, Ellison DW, Curran T, Phillips P, Goldman S, Packer R, Kun LE, Boyett JM, Gilbertson RJ (2010) A phase I pharmacokinetic trial of sonic hedgehog (SHH) antagonist GDC-0449 in pediatric patients with recurrent or refractory medulloblastoma: A Pediatric Brain Tumor Consortium study (PBTC 25). J Clin Oncol 2010 ASCO Annual Meeting Proceedings Vol 28, No 18_suppl (June 20 Supplement) Abstr: CRA9501

Goodrich LV, Scott MP (1998) Hedgehog and patched in neural development and disease. Neuron 21:1243–1257

Goodrich LV, Milenkovic L, Higgins KM, Scott MP (1997) Altered neural cell fates and medulloblastoma in mouse patched mutants. Science 277:1109–1113

Gorlin RJ (1987) Nevoid basal-cell carcinoma syndrome. Medicine (Baltimore) 66:98–113

Hahn H, Wicking C, Zaphiropoulous PG, Gailani MR, Shanley S, Chidambaram A, Vorechovsky I, Holmberg E, Unden AB, Gillies S, Negus K, Smyth I, Pressman C, Leffell DJ, Gerrard B, Goldstein AM, Dean M, Toftgard R, Chenevix-Trench G, Wainwright B, Bale AE (1996) Mutations of the human homolog of Drosophila patched in the nevoid basal cell carcinoma syndrome. Cell 85:841–851

Hellemans J, Coucke PJ, Giedion A, De Paepe A, Kramer P, Beemer F, Mortier GR (2003) Homozygous mutations in IHH cause acrocapitofemoral dysplasia, an autosomal recessive disorder with cone-shaped epiphyses in hands and hips. Am J Hum Genet 72:1040–1046

Huangfu D, Liu A, Rakeman AS, Murcia NS, Niswander L, Anderson KV (2003) Hedgehog signalling in the mouse requires intraflagellar transport proteins. Nature 426:83–87

Johnson RL, Rothman AL, Xie J, Goodrich LV, Bare JW, Bonifas JM, Quinn AG, Myers RM, Cox DR, Epstein EH Jr, Scott MP (1996) Human homolog of patched, a candidate gene for the basal cell nevus syndrome. Science 272:1668–1671

Kamachi Y, Kondoh H (1993) Overlapping positive and negative regulatory elements determine lens-specific activity of the delta 1-crystallin enhancer. Mol Cell Biol 13:5206–5215

Keeler RF (1972) Effect of natural teratogens in poisonous plants on fetal development in domestic animals. Adv Exp Med Biol 27:107–125

Kimonis VE, Goldstein AM, Pastakia B, Yang ML, Kase R, DiGiovanna JJ, Bale AE, Bale SJ (1997) Clinical manifestations in 105 persons with nevoid basal cell carcinoma syndrome. Am J Med Genet 69:299–308

Kimura H, Ng JMY, Curran T (2008) Transient inhibition of the Hedgehog pathway in young mice causes permanent defects in bone structure. Cancer Cell 13:249–260

Kinzler KW, Bigner SH, Bigner DD, Trent JM, Law ML, O'Brien SJ, Wong AJ, Vogelstein B (1987) Identification of an amplified, highly expressed gene in a human glioma. Science 236:70–73

Lee Y, Kawagoe R, Sasai K, Li Y, Russell HR, Curran T, McKinnon PJ (2007) Loss of suppressor-of-fused function promotes tumorigenesis. Oncogene 26:6442–6447

Li FP, Fraumeni JF (1969) Soft-tissue sarcomas, breast cancer, and other neoplasms A familial syndrome? Ann Intern Med 71:747–752

Louis DN, Ohgaki H, Wiestler OD, Cevenee WK, Burger PC, Jouvet A, Scheithauer BW, Kleihues P (2007) The 2007 WHO classification of tumors of the central nervous system. Acta Neuropathol 114:97–109

Lum L, Beachy PA (2004) The Hedgehog response network: sensors, switches, and routers. Science 304:1755–1759

Maeda Y, Nakamura E, Nguyen MT, Suva LJ, Swain FL, Razzaque MS, Mackem S, Lanske B (2007) Indian Hedgehog produced by postnatal chondrocytes is essential for maintaining a growth plate and trabecular bone. Proc Natl Acad Sci USA 104:6382–6387

Mas C, Altaba AR (2010) Small molecule modulation of HH-GLI signaling: current leads, trials and tribulations. Biochem Pharmacol 80:712–723

McMahon AP, Ingham PW, Tabin CJ (2003) Developmental roles and clinical significance of hedgehog signaling. Curr Top Dev Biol 53:1–114

Murone M, Rosenthal A, de Sauvage FJ (1999) Sonic hedgehog signaling by the patched-smoothened receptor complex. Curr Biol 9:76–84

Pietsch T, Waha A, Koch A, Kraus J, Albrecht S, Tonn J, Sorensen N, Berthold F, Henk B, Schmandt N, Wolf HK, von Deimling A, Wainwright B, Chenevix-Trench G, Wiestler OD, Wicking C (1997) Medulloblastomas of the desmoplastic variant carry mutations of the human homologue of Drosophila patched. Cancer Res 57:2085–2088

Pomeroy SL, Tamayo P, Gaasenbeek M, Sturla LM, Angelo M, McLaughlin ME, Kim JY, Goumnerova LC, Black PM, Lau C, Allen JC, Zagzag D, Olson JM, Curran T, Wetmore C, Biegel JA, Poggio T, Mukherjee S, Rifkin R, Califano A, Stolovitzky G, Louis DN, Mesirov JP, Lander ES, Golub TR (2002) Prediction of central nervous system embryonal tumour outcome based on gene expression. Nature 415:436–442

Raffel C, Jenkins RB, Frederick L, Hebrink D, Alderete B, Fults DW, James CD (1997) Sporadic medulloblastomas contain PTCH mutations. Cancer Res 57:842–845

Robarge KD, Brunton SA, Castanedo GM, Cui Y, Dina MS, Goldsmith R, Gould SE, Guichert O, Gunzner JL, Halladay J, Jia W, Khojasteh C, Koehler MF, Kotkow K, La H, Lalonde RL, Lau K, Lee L, Marshall D, Marsters JC Jr, Murray LJ, Qian C, Rubin LL, Salphati L, Stanley MS, Stibbard JH, Sutherlin DP, Ubhayaker S, Wang S, Wong S, Xie M (2009) GDC-0449-a potent inhibitor of the hedgehog pathway. Bioorg Med Chem Lett 19:5576–5581

Rohatgi R, Milenkovic L, Scott MP (2007) Patched1 regulates hedgehog signaling at the primary cilium. Science 317:372–376

Romer JT, Kimura H, Magdaleno S, Sasai K, Fuller C, Baines H, Connelly M, Stewart CF, Gould S, Rubin LL, Curran T (2004) Suppression of the Shh pathway using a small molecule inhibitor eliminates medulloblastoma in Ptc1(+/−)p53(−/−) mice. Cancer Cell 6:229–240

Rubin LL, de Sauvage FJ (2006) Targeting the Hedgehog pathway in cancer. Nat Rev Drug Discov 5:1026–1033

Sasai K, Romer JT, Lee Y, Finkelstein D, Fuller C, McKinnon PJ, Curran T (2006) Shh pathway activity is down-regulated in cultured medulloblastoma cells: implications for preclinical studies. Cancer Res 66:4215–4222

Sasaki H, Hui C, Nakafuku M, Kondoh H (1997) A binding site for Gli proteins is essential for HNF-3beta floor plate enhancer activity in transgenics and can respond to Shh in vitro. Development 124:1313–1322

St-Jacques B, Hammerschmidt M, McMahon AP (1999) Indian hedgehog signaling regulates proliferation and differentiation of chondrocytes and is essential for bone formation. Genes Dev 13:2072–2086

Taipale J, Cooper MK, Maiti T, Beachy PA (2002) Patched acts catalytically to suppress the activity of Smoothened. Nature 418:892–897

Taylor MD, Liu L, Raffel C, Hui CC, Mainprize TG, Zhang X, Agatep R, Chiappa S, Gao L, Lowrance A, Hao A, Goldstein AM, Stavrou T, Scherer SW, Dura WT, Wainwright B, Squire

JA, Rutka JT, Hogg D (2002) Mutations in SUFU predispose to medulloblastoma. Nat Genet 31:306–310

Thompson MC, Fuller C, Hogg TL, Dalton J, Finkelstein D, Lau CC, Chintagumpala M, Adesina A, Ashley DM, Kellie SJ, Taylor MD, Curran T, Gajjar A, Gilbertson RJ (2006) Genomics identifies medulloblastoma subgroups that are enriched for specific genetic alterations. J Clin Oncol 24:1924–1931

Von Hoff DD, LoRusso PM, Rudin CM, Reddy JC, Yauch RL, Tibes R, Weiss GJ, Borad MJ, Hann CL, Brahmer JR, Mackey HM, Lum BL, Darbonne WC, Marsters JC Jr, De Sauvage FJ, Low JA (2009) Inhibition of the hedgehog pathway in advanced basal-cell carcinoma. New Engl J Med 61:1164–1172

Wetmore C, Eberhart DE, Curran T (2000) The normal patched allele is expressed in medulloblastomas from mice with heterozygous germ-line mutation of patched. Cancer Res 60:2239–2246

Wetmore C, Eberhart DE, Curran T (2001) Loss of p53 but not ARF accelerates medulloblastoma in mice heterozygous for patched. Cancer Res 61:513–516

Williams JA, Guicherit OM, Zaharian BI, Xu Y, Chai L, Wichterle H, Kon C, Gatchalian C, Porter JA, Rubin LL, Wang FY (2003) Identification of a small molecule inhibitor of the hedgehog signaling pathway: effects on basal cell carcinoma-like lesions. Proc Natl Acad Sci USA 100:4616–4621

Xie J, Johnson RL, Zhang X, Bare JW, Waldman FM, Cogen PH, Menon AG, Warren RS, Chen LC, Scott MP, Epstein EH Jr (1997) Mutations of the PATCHED gene in several types of sporadic extracutaneous tumors. Cancer Res 57:2369–2372

Yauch RL, Gould SE, Scales SJ, Tang T, Tian H, Ahn CP, Marshall D, Fu L, Januario T, Kallop D, Bazan JF, Kan Z, Seshagiri S, Hann CL, Gould SE, Low JA, Rudin CM, de Sauvage FJ (2008) A paracrine requirement for hedgehog signalling in cancer. Nature 455:406–410

Yauch RL, Dijkgraaf GJP, Alicke B, Januario T, Ahn CP, Holcomb T, Pujara K, Stinson J, Callahan CA, Tang T, Nannini-Pepe M, Kotkow K, Marsters JC, Rubin LL, de Sauvage FJ (2009) Smoothened mutation confers resistance to a Hedgehog pathway inhibitor in medulloblastoma. Science 326:572–574

Primary Cilia as Switches in Brain Development and Cancer

Young-Goo Han and Arturo Alvarez-Buylla

Abstract The primary cilium plays key roles in regulating the cell cycle. This specialized organelle grows from the basal body, a modified centriole that during mitosis functions as part of the centrosome organizing the mitotic spindle. The primary cilium is essential for neural stem cell specification and neural progenitor expansion. Growing evidence indicates that several brain tumors arise from progenitor or stem cells. Recent work has demonstrated that the primary cilium is a sensory organelle that concentrates receptors and second messenger components for the hedgehog (Hh) signaling pathway. Several types of cancers show oncogenic activation of Hh signaling, yet the role for primary cilia in cancer has only recently been addressed. Interestingly, the primary cilium appears to have dual and opposing roles in cancer; depending on the underlying oncogenic event, primary cilia can either promote or prevent tumor formation. Here, we suggest that primary cilia function as ON or OFF switches to regulate cell proliferation and as such could have important implications in the diagnosis and treatment of cancer.

1 Introduction

The primary cilium is a cylindrical organelle that protrudes from the surface of most vertebrate cells. Primary cilia are covered by an extension of the cell membrane and contain a characteristic microtubular structure called the axoneme, a ring of nine microtubule doublets (9+0) that runs longitudinally through the organelle. This arrangement is different from that of typical motile cilia, which contain an additional pair of microtubules in the center (9+2). The axonemes of primary cilia grow from the basal body, a modified mother centriole, which converts into the centrosome to organize the mitotic spindle during cell division. Thus, ciliogenesis

A. Alvarez-Buylla (✉)

Department of Neurological Surgery, The Eli and Edythe Broad Center of Regeneration Medicine and Stem Cell Research, University of California, San Francisco, CA 94143, USA

e-mail: abuylla@stemcell.ucsf.edu

T. Curran and Y. Christen (eds.), *Two Faces of Evil: Cancer and Neurodegeneration*,
Research and Perspectives in Alzheimer's Disease, DOI 10.1007/978-3-642-16602-0_6,
© Springer-Verlag Berlin Heidelberg 2011

is tightly regulated throughout cell cycle; proliferating cells disassemble cilia before mitosis and reassemble them during interphase (Rieder et al. 1979; Tucker et al. 1979). Consistent with this regulation, Aurora A kinase, which regulates mitotic entry, triggers ciliary disassembly (Pugacheva et al. 2007), whereas cyclin-dependent kinase inhibitor 2B positively regulates ciliogenesis (Kim et al. 2010). In addition, the primary cilium is an essential organizing center for multiple signaling pathways that critically regulate cell proliferation and differentiation, including Hedgehog (Hh), platelet-derived growth factor (PDGF), Wnt, and G-protein coupled receptor signaling pathways (Goetz and Anderson 2010). Thus, primary cilia could relay signals from the microenvironment directly to the modified centriole. Consistent with these reciprocal interconnections between primary cilia and cell cycle progression, primary cilia play critical roles in development and tumorigenesis, as discussed below (Goetz and Anderson 2010; Han and Alvarez-Buylla 2010). Defective cilia or basal body function is associated with a number of human genetic syndromes including Bardet Biedl, Joubert, Meckel-Gruber, and MORM syndromes (Gerdes et al. 2009). These ciliopathies are complex disorders resulting in multi-organ defects including cystic kidneys, retinal degeneration, polydactyly, obesity, cognitive impairment and ataxia. These pleiotropic phenotypes (each disease has its own phenotype) highlight the crucial roles of primary cilia in many aspects of human development and physiology.

2   Ciliogenesis

Although the primary cilium is continuous with the cytoplasm, cilia are discrete subcellular compartments containing components distinct from the rest of the cell. This distinction is generated and maintained by intraflagellar transport (IFT), a bidirectional trafficking mechanism along the axonemal microtubules (Pedersen and Rosenbaum 2008). In IFT, kinesin-II, a heterotrimeric motor complex, moves ciliary components from the basal body into the cilium and a cytoplasmic dynein motor complex brings components back to the basal body. The multiprotein adaptor complexes IFTA and IFTB are also essential for retrograde and antero-grade transport, respectively. Cilia are highly evolutionarily conserved organelles, and the process of IFT is also broadly conserved: mutations disrupting IFT cause defective ciliogenesis in organisms ranging from the unicellular organisms to humans. Important insights into the function of primary cilia came from analyses of IFT mutants that have defective cilia. Like the inside of cilia, the ciliary membrane contains unique lipid and protein compositions. This unique ciliary membrane does not form by passive evagination pushed by the elongating axoneme but rather is driven by specific Golgi-drived vesicles and the recycling endosome pathway (Kim et al. 2010; Rohatgi and Snell 2010). Interestingly, proteins encoded by seven of the twelve genes mutated in Bardet-Biedl syndrome (BBS) form a protein complex that interacts with Rab8 and promotes ciliary membrane biogenesis (Nachury et al. 2007). A recent study showed that the coordination of axoneme and ciliary membrane assemblies depends on cell cycle-related kinase and Broad-minded, a novel TBC domain-containing protein (Ko et al. 2010).

Ciliogenesis is also regulated by extracellular signaling pathways. Studies in zebrafish and frog have shown that Fibroblast growth factor (Fgf) signaling regulates cilia length; knockdown of *fgf4, fgf8, fgf24, or fgf receptor 1* resulted in reduced cilia length (Neugebauer et al. 2009; Yamauchi et al. 2009). In these studies, only motile cilia were examined, and it will be important to see if fgf signaling regulates primary cilia and ciliogenesis in the mouse. Inositol signaling pathways also regulate ciliogenesis. Knockdown of inositol 1,3,4,5,6-pentakisphosphate 2-kinase reduces motile cilia length in zebrafish (Sarmah et al. 2007). In humans, inositol polyphosphate 5-phosphatase is mutated in two types of ciliopathies, Joubert and MORM syndromes, and these mutations cause primary cilia to become unstable upon mitogenic stimulus (Bielas et al. 2009; Jacoby et al. 2009).

3   Primary cilia and Hh signaling

Studies in vertebrates during the past decade have demonstrated that the primary cilium is essential for Hh signaling. Hh is a secreted morphogen that controls the patterning, development, and homeostasis of many animal tissues by regulating cell proliferation and differentiation (Varjosalo and Taipale 2008). Decreased Hh activity causes severe developmental defects, including holoprosencephaly, poly-dactyly and skeletal malformation. Uncontrolled activation of Hh signaling is associated with many cancers, including basal cell carcinoma, rhabdomyosarcoma and medulloblastoma (Varjosalo and Taipale 2008). The first clue linking primary cilia to Hh signaling came from forward genetic studies in mice (Huangfu et al. 2003). Mouse mutants with defective IFT show developmental defects similar to Hh signaling mutants, and epigenetic analyses placed primary cilia downstream of the Hh receptor Patched1 (Ptch1) and the transmembrane protein Smoothened (Smo) but upstream of the GLI-Kruppel family member transcription factors (Gli1-3). Subsequent studies showed that Smo moves into the primary cilium to activate the Hh signaling pathway (Corbit et al. 2005). In the absence of Hh ligands, Ptch1 localizes to primary cilia, inhibiting ciliary localization of Smo. Hh binding to Ptch1 results in this receptor moving out of the cilia and accumulation of Smo into the cilium (Rohatgi et al. 2007). Importantly, the Gli transcription factors also localize to primary cilia, and loss of cilia disrupts both the formation of Gli2 activator in the presence of Hh ligand and the production of Gli3 repressor in the absence of Hh (Haycraft et al. 2005; Huangfu and Anderson 2005; Liu et al. 2005; May et al. 2005). Thus, primary cilia are essential to both turning on and turning off the Hh signaling pathway (Fig. 1).

Additional evidence indicates that primary cilia actively regulate the Hh signaling pathway rather than just concentrating signaling components to facilitate their interactions. In contrast to the complete loss of primary cilia, which leads to loss of both Gli activators and repressors, partial defects in ciliary structure lead to distinct defects in the Hh signaling pathways. Arl13b, a small GTPase of the Arf/Arl family, localizes to primary cilia. Loss of Arl13b causes specific axonemal defects resulting in the opening of B microtubules of doublet microtubules and constitutive activation of Gli activators at low levels without disrupting Gli3 repressor activity (Caspary et al. 2007). Mutations affecting the components of the IFTA complex (required for retrograde transport) lead to short cilia with a bulge at the ciliary tip. These IFTA mutants show ectopic and constitutive activation of the Hh pathway (Tran et al. 2008; Cortellino et al. 2009). Lastly, loss of broad-minded leads to detachment of the ciliary membrane from the axoneme and failure of cells to respond to high Hh levels (Ko et al. 2010).

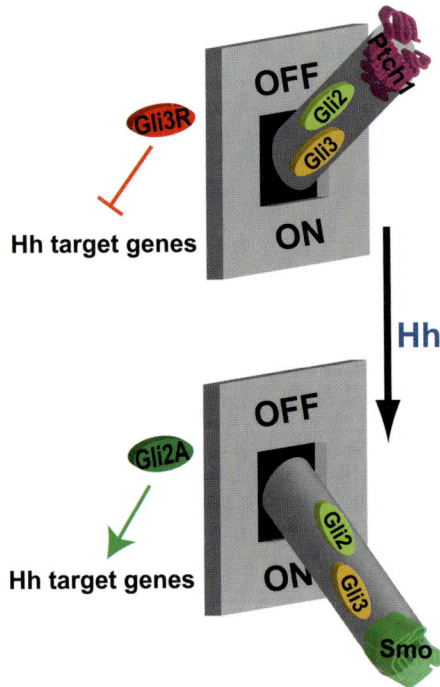

Fig. 1 Primary cilia as switches in the Hh signaling pathway. In the absence of Hh, Ptch1 localizes to the primary cilium and inhibits ciliary accumulation of Smo. Without activated Smo in primary cilia, Gli3 transcription factor is constitutively processed to the repressor form (Gli3R), which represses the expression of Hh target genes. Primary cilia are essential to produce Gli3R, thus keeping the SWITCH OFF for the Hh target genes. Binding of Hh to Ptch1 allows ciliary accumulation of Smo, promoting Gli2 activator formation and inhibiting Gli3R formation. Primary cilia are required for this Smo activity to SWITCH ON the Hh signaling pathway

4 Primary Cilia in the Specification of Neural Progenitors

The brain consists of discrete domains containing a variety of cell types that carry out unique functions. The production of distinct cell types is achieved during development through regional specification of proliferating neural progenitors. The primary cilium critically regulates neural progenitor specification and proliferation through its dual role in the Hh signaling pathways. Recent studies have shown that defective cilia disrupt proper specification of neural progenitors along the dorsal-ventral and rostral-caudal axes (Willaredt et al. 2008; Gorivodsky et al. 2009; Stottmann et al. 2009). Mutations in IFT components, either IFT88 (Willaredt et al. 2008) or IFT139 (Stottmann et al. 2009), result in loss of dorsomedial telencephalic structures, dorsoventral patterning defects, and lack of clear distinction between the telencephalon and diencephalon. Consistent with the essential role of primary cilia in the formation of Gli3 repressors, these phenotypes are very similar to defects observed in *Gli3* mutant mice (Theil et al. 1999; Tole et al. 2000; Fotaki et al. 2006). Primary cilia appear to play critical roles in brain patterning by producing Gli3 repressors, which likely restrict the expression of a set of genes for the proper specification of specific sets of neural progenitors.

During perinatal development, primary cilia are also essential to mediate Hh signaling for the proliferation of neural progenitors. Conditional removal of ciliogenic genes (either *Kif3a,* which encodes a subunit of Kinesin-II, or *stumpy,* a novel basal body protein) in neural progenitors using a *Nestin* or *hGFAP* promoter-driven Cre dramatically reduces the proliferation of granule neuron precursors (GNPs) in the developing dentate gyrus (Breunig et al. 2008; Han et al. 2008). During early postnatal life, GNPs transform into adult neural progenitors, which are required for the continual production of new neurons throughout life (Seri et al. 2001; Filippov et al., 2003; Fukuda et al. 2003). The dentate gyrus of ciliary mutants is not only small but also lacks the radial astrocytes that function as adult neural stem cells. The primary cilium is, therefore, essential for the transition from embryonic to adult neural progenitors in the hippocampus. Similar to the function of primary cilia in early brain patterning, its function in the specification of dentate gyrus progenitors appears to be primarily as a mediator of Hh signaling. Conditional ablation of Smo in the same neural progenitors results in a similar small dentate gyrus with no radial astrocytes (Han et al. 2008). In contrast, conditional activation of Hh signaling via expression of a constitutively active form of Smo (SmoM2) results in the expansion of GNPs. This expansion of GNPs also requires the primary cilia, as SmoM2 expression in the absence of cilia fails to expand GNPs (Han et al. 2008).

Another region of the brain where primary cilia play a key role in neural precursor proliferation is in the cerebellum. The Sonic Hedgehog isoform secreted by Purkinje neurons is essential for the proliferation of cerebellar GNPs. Consistent with the essential role of cilia in Hh signaling, removal of primary cilia in cerebellar GNPs results in severe hypoplasia and abnormal foliation (Chizhikov et al. 2007; Breunig et al. 2008; Spassky et al. 2008). Interestingly, while expression of SmoM2 in dentate gyrus GNPs does not induce tumors in the forebrain, expression of this same mutant protein in the cerebellar GNPs results in a fully penetrant medulloblastoma phenotype.

5 Primary Cilia and Tumorigenesis

While Hh signaling is essential for GNP proliferation, constitutive activation of this signaling in cerebellar GNPs can result in medulloblastoma (Gilbertson and Ellison 2008), the most common malignant brain tumor in children. Recently, we showed that primary cilia have surprising dual roles in this Hh signaling-driven tumorigenesis (Han et al. 2009). A tumor-derived mutant of Smo, SmoM2, is not oncogenic in GNPs in the hippocampus (see above), but its expression in cerebellar GNPs induces medulloblastoma formation. In these mice, SmoM2 constitutively localizes to the primary cilium and requires this organelle to induce tumorigenesis; concomitant removal of primary cilia in SmoM2-expressing cells completely blocks tumor formation. In contrast to SmoM2, a constitutively active form of GLI2 (GLI2ΔN) (Pasca di Magliano et al. 2006), which lacks the N-terminal repressor domain, fails to induce medulloblastoma. Surprisingly, however, removal of primary cilia allows GLI2ΔN to induce medulloblastoma, indicating that primary cilia function as a tumor suppressor in GLI2ΔN-expressing cells (Han et al. 2009). It remains to be determined how primary cilia suppress GLI2ΔN-driven medulloblastoma. One possibility is that removal of primary cilia disrupts the formation of the Gli3 repressor, which could counteract the oncogenic activity of GLI2ΔN. Thus, the primary cilium appears to function as an ON or OFF switch in tumorigenesis. Oncogenic mutations upstream of the primary cilium need this organelle to keep signaling ON and induce medulloblastoma. In contrast, mutations downstream of the cilia need to ablate the OFF signal. One way in which tumor cells may do this is by acquiring mutations that get rid of the primary cilium switch entirely. Interestingly, this is precisely what is observed in a subset of human tumors. Some human medulloblastomas have primary cilia whereas others do not. Notably, primary cilia are present almost exclusively in medulloblastomas with activations in HH or WNT signaling and absent in medulloblastomas with other molecular changes (Thompson et al. 2006; Han et al. 2009) (Fig. 2).

Primary cilia may critically regulate the formation of other types of tumors. In fact, primary cilia also have a similar dual function in skin tumors driven by SmoM2 and GLI2ΔN (Wong et al. 2009). Moreover, the important role of cilia may not be restricted to Hh-driven tumorigenesis. The presence of cilia in human medulloblastomas driven by WNT activation, but its absence in other subtypes supports this hypothesis (Han et al. 2009). In addition, loss of cilia is a cardinal feature of human pancreatic ductal adenocarcinoma (PDAC), which shows Kras mutations in more than 90% of tumors (Seeley et al. 2009). Cilia are also absent in mouse models of PDAC driven by Kras. Notably, primary cilia are absent only in tumors but not in normal pancreatic tissues that express oncogenic Kras, suggesting

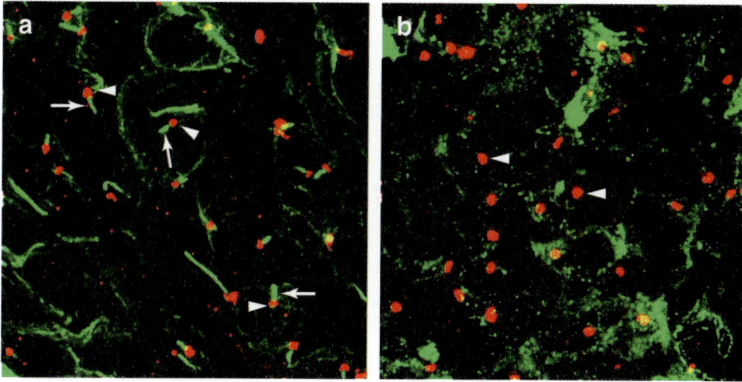

Fig. 2 Primary cilia in human medulloblastoma. Human medulloblastomas showing activation of either SHH or WNT signaling have primary cilia (**a**, *arrows*) whereas most of the others do not have primary cilia (**b**; Thompson et al. 2006; Han et al. 2009). Basal bodies are present in both cases. Human medulloblastoma sections are stained with an antibody against acetylated tubulin (*green, arrows*) to label primary cilia and an antibody against pericentrin (*red, arrowheads*) to label basal bodies

that the loss of cilia is specifically coupled to tumorigenesis (Seeley et al. 2009). Loss of cilia is also prevalent in clear cell renal cell carcinomas, which is frequently associated with germline mutations of the *von Hippel–Lindau* (*VHL*) tumor suppressor gene, but does not occur in papillary renal cell carcinomas (Schraml et al. 2009). VHL is a component of the E3 ubiquitin ligase complex that targets hypoxia-inducible factor for destruction, and recent studies show that VHL is required for ciliogenesis as well (Esteban et al. 2006; Lutz and Burk 2006; Thoma et al. 2007). It will be interesting to test if primary cilia have a role in oxygen sensing and differentially affect the formation of these two types of renal carcinomas. Lastly, defective ciliogenesis is also observed in human astrocytoma/glioblastoma cell lines (Moser et al. 2009).

6 Conclusion

The primary cilium is at the crossroads of the cell cycle and extracellular signaling pathways: the cell cycle governs ciliogenesis and cilia regulate cell cycle through their role as a signaling center. Most vertebrate cells have primary cilia, and tumor cells may therefore have primary cilia as a default. Loss of cilia in tumor cells may be a secondary consequence of having compromised cell cycle regulation. However, at least in Hh-signaling driven tumors, the presence or absence of cilia is critical for tumors to form. Here, we propose that the primary cilium functions like a switch turning ON or OFF signaling pathways critical for cell cycle progression. Under basal conditions, this switch is normally off, providing strong inhibitory signaling blocking cell cycle progression. For tumors that do not require signaling through the cilium, this organelle likely interferes with cell proliferation and these tumors simply eliminate it. In contrast, some tumors are likely to require intact cilia to transduce oncogenic signaling acting upstream of the cilia. It remains unknown whether primary cilia play a role in tumors that are not associated with oncogenic Hh signaling. Further elucidation of the molecular mechanism of ciliary function in Hh and other signaling pathways could also provide important therapeutics for cancer treatment. In particular, a full understanding of the basal inhibitory function of primary cilia in cell cycle progression may provide candidate molecules or targets to inhibit tumor growth. In addition, targeting the primary cilia in tumors that depend on this structure for oncogenic growth may interfere with the growth of these tumors. These studies will further reveal the potential of primary cilia as a diagnostic or prognostic tool for tumors and as a potential therapeutic target.

Acknowledgments Y.-G. H. was, in part, supported by Mark Linder/American Brain Tumor Association Fellowship. The work was supported by grants from the US National Institute of Health (NS28478 and HD32116) and a grant from the Goldhirsh foundation to A. A.-B. We thank R. Ihrie for comments on the manuscript.

References

Bielas SL, Silhavy JL, Brancati F, Kisseleva MV, Al-Gazali L, Sztriha L, Bayoumi RA, Zaki MS, Abdel-Aleem A, Rosti RO, Kayserili H, Swistun D, Scott LC, Bertini E, Boltshauser E, Fazzi E, Travaglini L, Field SJ, Gayral S, Jacoby M, Schurmans S, Dallapiccola B, Majerus PW, Valente EM, Gleeson JG (2009) Mutations in INPP5E, encoding inositol polyphosphate-5-phosphatase E, link phosphatidyl inositol signaling to the ciliopathies. Nat Genet 41:1032–1036
Breunig JJ, Sarkisian MR, Arellano JI, Morozov YM, Ayoub AE, Sojitra S, Wang B, Flavell RA, Rakic P, Town T (2008) Primary cilia regulate hippocampal neurogenesis by mediating sonic hedgehog signaling. Proc Natl Acad Sci USA 105:13127–13132
Caspary T, Larkins CE, Anderson KV (2007) The graded response to Sonic Hedgehog depends on cilia architecture. Dev Cell 12:767–778
Chizhikov VV, Davenport J, Zhang Q, Shih EK, Cabello OA, Fuchs JL, Yoder BK, Millen KJ (2007) Cilia proteins control cerebellar morphogenesis by promoting expansion of the granule progenitor pool. J Neurosci 27:9780–9789
Corbit KC, Aanstad P, Singla V, Norman AR, Stainier DY, Reiter JF (2005) Vertebrate Smoothened functions at the primary cilium. Nature 437:1018–1021
Cortellino S, Wang C, Wang B, Bassi MR, Caretti E, Champeval D, Calmont A, Jarnik M, Burch J, Zaret KS, Larue L, Bellacosa A (2009) Defective ciliogenesis, embryonic lethality and severe impairment of the Sonic Hedgehog pathway caused by inactivation of the mouse complex A intraflagellar transport gene Ift122/Wdr10, partially overlapping with the DNA repair gene Med1/Mbd4. Dev Biol 325:225–237
Esteban MA, Harten SK, Tran MG, Maxwell PH (2006) Formation of primary cilia in the renal epithelium is regulated by the von Hippel-Lindau tumor suppressor protein. J Am Soc Nephrol 17:1801–1806
Filippov V, Kronenberg G, Pivneva T, Reuter K, Steiner B, Wang LP, Yamaguchi M, Kettenmann H, Kempermann G (2003) Subpopulation of nestin-expressing progenitor cells in the adult murine hippocampus shows electrophysiological and morphological characteristics of astrocytes. Mol Cell Neurosci 23:373–382

Fotaki V, Yu T, Zaki PA, Mason JO, Price DJ (2006) Abnormal positioning of diencephalic cell types in neocortical tissue in the dorsal telencephalon of mice lacking functional Gli3. J Neurosci 26:9282–9292

Fukuda S, Kato F, Tozuka Y, Yamaguchi M, Miyamoto Y, Hisatsune T (2003) Two distinct subpopulations of nestin-positive cells in adult mouse dentate gyrus. J Neurosci 23:9357–9366

Gerdes JM, Davis EE, Katsanis N (2009) The vertebrate primary cilium in development, homeostasis, and disease. Cell 137:32–45

Gilbertson RJ, Ellison DW (2008) The origins of medulloblastoma subtypes. Annu Rev Pathol 3:341–365

Goetz SC, Anderson KV (2010) The primary cilium: a signalling centre during vertebrate development. Nat Rev Genet 11:331–344

Gorivodsky M, Mukhopadhyay M, Wilsch-Braeuninger M, Phillips M, Teufel A, Kim C, Malik N, Huttner W, Westphal H (2009) Intraflagellar transport protein 172 is essential for primary cilia formation and plays a vital role in patterning the mammalian brain. Dev Biol 325:24–32

Han YG, Alvarez-Buylla A (2010) Role of primary cilia in brain development and cancer. Curr Opin Neurobiol 20:58–67

Han YG, Spassky N, Romaguera-Ros M, Garcia-Verdugo JM, Aguilar A, Schneider-Maunoury S, Alvarez-Buylla A (2008) Hedgehog signaling and primary cilia are required for the formation of adult neural stem cells. Nat Neurosci 11:277–284

Han YG, Kim HJ, Dlugosz AA, Ellison DW, Gilbertson RJ, Alvarez-Buylla A (2009) Dual and opposing roles of primary cilia in medulloblastoma development. Nat Med 15:1062–1065

Haycraft CJ, Banizs B, Aydin-Son Y, Zhang Q, Michaud EJ, Yoder BK (2005) Gli2 and Gli3 localize to cilia and require the intraflagellar transport protein polaris for processing and function. PLoS Genet 1:e53

Huangfu D, Anderson KV (2005) Cilia and Hedgehog responsiveness in the mouse. Proc Natl Acad Sci USA 102:11325–11330

Huangfu D, Liu A, Rakeman AS, Murcia NS, Niswander L, Anderson KV (2003) Hedgehog signalling in the mouse requires intraflagellar transport proteins. Nature 426:83–87

Jacoby M, Cox JJ, Gayral S, Hampshire DJ, Ayub M, Blockmans M, Pernot E, Kisseleva MV, Compere P, Schiffmann SN, Gergely F, Riley JH, Perez-Morga D, Woods CG, Schurmans S (2009) INPP5E mutations cause primary cilium signaling defects, ciliary instability and ciliopathies in human and mouse. Nat Genet 41:1027–1031

Kim J, Lee JE, Heynen-Genel S, Suyama E, Ono K, Lee K, Ideker T, Aza-Blanc P, Gleeson JG (2010) Functional genomic screen for modulators of ciliogenesis and cilium length. Nature 464:1048–1051

Ko HW, Norman RX, Tran J, Fuller KP, Fukuda M, Eggenschwiler JT (2010) Broad-minded links cell cycle-related kinase to cilia assembly and hedgehog signal transduction. Dev Cell 18:237–247

Liu A, Wang B, Niswander LA (2005) Mouse intraflagellar transport proteins regulate both the activator and repressor functions of Gli transcription factors. Development 132:3103–3111

Lutz MS, Burk RD (2006) Primary cilium formation requires von hippel-lindau gene function in renal-derived cells. Cancer Res 66:6903–6907

May SR, Ashique AM, Karlen M, Wang B, Shen Y, Zarbalis K, Reiter J, Ericson J, Peterson AS (2005) Loss of the retrograde motor for IFT disrupts localization of Smo to cilia and prevents the expression of both activator and repressor functions of Gli. Dev Biol 287:378–389

Moser JJ, Fritzler MJ, Rattner JB (2009) Primary ciliogenesis defects are associated with human astrocytoma/glioblastoma cells. BMC Cancer 9:448

Nachury MV, Loktev AV, Zhang Q, Westlake CJ, Peranen J, Merdes A, Slusarski DC, Scheller RH, Bazan JF, Sheffield VC, Jackson PK (2007) A core complex of BBS proteins cooperates with the GTPase Rab8 to promote ciliary membrane biogenesis. Cell 129:1201–1213

Neugebauer JM, Amack JD, Peterson AG, Bisgrove BW, Yost HJ (2009) FGF signalling during embryo development regulates cilia length in diverse epithelia. Nature 458:651–654

Pasca di Magliano M, Sekine S, Ermilov A, Ferris J, Dlugosz AA, Hebrok M (2006) Hedgehog/
  Ras interactions regulate early stages of pancreatic cancer. Genes Dev 20:3161–3173
Pedersen LB, Rosenbaum JL (2008) Intraflagellar transport (IFT) role in ciliary assembly,
  resorption and signalling. Curr Top Dev Biol 85:23–61
Pugacheva EN, Jablonski SA, Hartman TR, Henske EP, Golemis EA (2007) HEF1-dependent
  Aurora A activation induces disassembly of the primary cilium. Cell 129:1351–1363
Rieder CL, Jensen CG, Jensen LC (1979) The resorption of primary cilia during mitosis in a
  vertebrate (PtK1) cell line. J Ultrastruct Res 68:173–185
Rohatgi R, Snell WJ (2010) The ciliary membrane. Curr Opin Cell Biol 22(4):541–546
Rohatgi R, Milenkovic L, Scott MP (2007) Patched1 regulates hedgehog signaling at the primary
  cilium. Science 317:372–376
Sarmah B, Winfrey VP, Olson GE, Appel B, Wente SR (2007) A role for the inositol kinase Ipk1 in
  ciliary beating and length maintenance. Proc Natl Acad Sci USA 104:19843–19848
Schraml P, Frew IJ, Thoma CR, Boysen G, Struckmann K, Krek W, Moch H (2009) Sporadic clear
  cell renal cell carcinoma but not the papillary type is characterized by severely reduced
  frequency of primary cilia. Mod Pathol 22:31–36
Seeley ES, Carriere C, Goetze T, Longnecker DS, Korc M (2009) Pancreatic cancer and precursor
  pancreatic intraepithelial neoplasia lesions are devoid of primary cilia. Cancer Res 69:422–430
Seri B, Garcia-Verdugo JM, McEwen BS, Alvarez-Buylla A (2001) Astrocytes give rise to new
  neurons in the adult mammalian hippocampus. J Neurosci 21:7153–7160
Spassky N, Han YG, Aguilar A, Strehl L, Besse L, Laclef C, Ros MR, Garcia-Verdugo JM,
  Alvarez-Buylla A (2008) Primary cilia are required for cerebellar development and Shh-
  dependent expansion of progenitor pool. Dev Biol 317:246–259
Stottmann RW, Tran PV, Turbe-Doan A, Beier DR (2009) Ttc21b is required to restrict sonic
  hedgehog activity in the developing mouse forebrain. Dev Biol 335:166–178
Theil T, Alvarez-Bolado G, Walter A, Ruther U (1999) Gli3 is required for Emx gene expression
  during dorsal telencephalon development. Development 126:3561–3571
Thoma CR, Frew IJ, Hoerner CR, Montani M, Moch H, Krek W (2007) pVHL and GSK3beta are
  components of a primary cilium-maintenance signalling network. Nat Cell Biol 9:588–595
Thompson MC, Fuller C, Hogg TL, Dalton J, Finkelstein D, Lau CC, Chintagumpala M, Adesina
  A, Ashley DM, Kellie SJ, Taylor MD, Curran T, Gajjar A, Gilbertson RJ (2006) Genomics
  identifies medulloblastoma subgroups that are enriched for specific genetic alterations. J Clin
  Oncol 24:1924–1931
Tole S, Ragsdale CW, Grove EA (2000) Dorsoventral patterning of the telencephalon is disrupted
  in the mouse mutant extra-toes(J). Dev Biol 217:254–265
Tran PV, Haycraft CJ, Besschetnova TY, Turbe-Doan A, Stottmann RW, Herron BJ, Chesebro
  AL, Qiu H, Scherz PJ, Shah JV, Yoder BK, Beier DR (2008) THM1 negatively modulates
  mouse sonic hedgehog signal transduction and affects retrograde intraflagellar transport in
  cilia. Nat Genet 40:403–410
Tucker RW, Pardee AB, Fujiwara K (1979) Centriole ciliation is related to quiescence and DNA
  synthesis in 3T3 cells. Cell 17:527–535
Varjosalo M, Taipale J (2008) Hedgehog: functions and mechanisms. Genes Dev 22:2454–2472
Willaredt MA, Hasenpusch-Theil K, Gardner HA, Kitanovic I, Hirschfeld-Warneken VC, Gojak
  CP, Gorgas K, Bradford CL, Spatz J, Wolfl S, Theil T, Tucker KL (2008) A crucial role for
  primary cilia in cortical morphogenesis. J Neurosci 28:12887–12900
Wong SY, Seol AD, So PL, Ermilov AN, Bichakjian CK, Epstein EH Jr, Dlugosz AA, Reiter JF
  (2009) Primary cilia can both mediate and suppress Hedgehog pathway-dependent tumorigenesis. Nat Med 15:1055–1061
Yamauchi H, Miyakawa N, Miyake A, Itoh N (2009) Fgf4 is required for left-right patterning of
  visceral organs in zebrafish. Dev Biol 332:177–185

Nervous System Aging, Degeneration, and the p53 Family

Freda D. Miller and David R. Kaplan

Abstract Over the past decade, evidence has emerged implicating the p53 family members p73 and p63 in two functions that likely play key roles in maintaining long-term nervous system integrity. First, p73, acting as a truncated, dominant-inhibitory $\Delta Np73$ isoform, promotes neuronal survival and maintains neuronal longevity in the face of normal ongoing "wear and tear" and following injuries or insults. Second, p63 promotes the survival of neural precursors, thereby regulating the establishment and, potentially, maintenance of adult neural stem cell (NSC) pools. On the basis of these findings, we propose that p73 and p63 act to maintain neurons and NSC pools over an animal's life span and, in so doing, prevent premature nervous system aging and neurodegeneration. Here, we will review the data supporting the idea that p73 and p63 perform these longevity functions.

1 Introduction

The cellular mechanisms that regulate cell survival and maintenance in the aged or degenerating nervous system are only poorly understood, in spite of their clinical relevance. Much of what we know about cell death comes from studies of naturally occurring developmental cell death, which occurs largely via apoptosis. However, the relevance of these relatively well-understood apoptotic mechanisms to aging or neurodegeneration is still an open question. For example, it is unclear how important apoptosis is in neurodegenerative disorders such as Alzheimer's disease (AD), and how any apoptosis that does occur is similar to developmental cell death (for example, see Dusart et al. 2006). Moreover, the pathology observed in many neurodegenerative disorders indicates that neurons accumulate intracellular protein

F.D. Miller (✉)
Developmental and Stem Cell Biology and Departments of Molecular Genetics and Physiology, University of Toronto, Toronto, ON, Canada M5G 1L7

T. Curran and Y. Christen (eds.), *Two Faces of Evil: Cancer and Neurodegeneration*, Research and Perspectives in Alzheimer's Disease, DOI 10.1007/978-3-642-16602-0_7, © Springer-Verlag Berlin Heidelberg 2011

aggregates and/or cellular damage that leads to their slow degeneration and death, as opposed to a rapid and controlled apoptotic suicide. Instead, increasing evidence indicates that, in the aged or degenerating nervous system, prolonged exposure to relatively minor insults might lead to cumulative neuronal oxidative stress and/or DNA damage, which in turn causes neurons to senesce, re-enter the cell cycle and/or atrophy and ultimately degenerate (Lin and Beal 2006; Herrup and Yang 2007).

One family of proteins that is essential for an appropriate response to oxidative stress and/or DNA damage is the p53 family. The p53 tumor suppressor is a transcription factor that is mutated or inactivated in >50% of human cancers (Vogelstein et al. 2000); in response to cellular stresses like DNA damage, it can either activate or repress a wide array of gene targets, which in turn can regulate apoptosis, proliferation, and DNA repair. For many years, p53 was thought to be "alone" in the genome, but in the late 1990s, two new genes with high homology to p53 were discovered. These genes, p63 (Yang et al. 1998) and p73 (Kaghad et al. 1997; Jost et al. 1997), encode transcription factors with striking similarity to p53 (Fig. 1); they bind to DNA, transactivate at least some p53 target genes, and induce apoptosis (reviewed in Jacobs et al. 2006). However, unlike p53, which is predominantly expressed as a single, full-length protein isoform, many different isoforms are generated from the p73 and p63 genes. Of particular importance, alternative promoter usage generates p73 and p63 N-terminal truncated isoforms lacking the transactivation domain (ΔNp73 and ΔNp63; Fig. 1) that are naturally occurring, dominant-inhibitory family members.

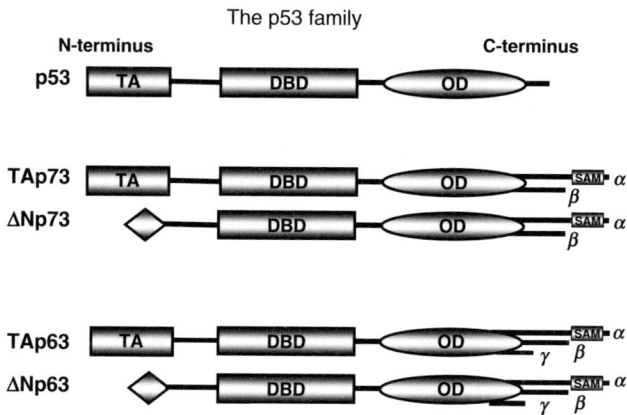

Fig. 1 Structure of the p53 family members. P53 is a transcription factor that is comprised of a transactivation domain (TA), a DNA-binding domain (DBD) and an oligomerization domain (OD). The other p53 family members, p73 and p63, exist as two major classes of isoforms, the full-length TA isoforms (TAp73, TAp63) and the N-terminal truncated ΔN isoforms (ΔNp73, ΔNp63) that lack the transactivation domain. These N-terminal variants are predominantly generated by alternative promoter usage. There are also multiple C-terminal variants of p63 and p73 that are generated by alternative splicing. Some, but not all, of these are shown in this schematic. For a more complete description of these variants, see Murray-Zmijewski et al. (2006)

2   The p53 Family Regulates Neuronal Survival and Longevity

The first indication that the p53 family was important for neural cell survival came from work on p53 itself, showing that it was increased in injured neurons and that this increase caused neuronal apoptosis (reviewed in Jacobs et al. 2006). At about the same time, we obtained evidence implicating p53 in naturally occurring sympathetic neuron death. Sympathetic neurons are over-produced during development, and their survival is determined by competition for limited amounts of target-derived nerve growth factor (NGF; reviewed in Kaplan and Miller 2000). NGF binds to its cognate TrkA tyrosine kinase receptor on sympathetic axon terminals, which transduces a retrograde survival signal down the axon to the cell body. Those neurons that are unable to sequester sufficient NGF die as a consequence of the lack of TrkA survival signaling and coincident apoptotic signaling by the p75 neurotrophin receptor, a member of the fas death receptor family. In this regard, we showed that p53 was an essential proapoptotic protein during this developmental death period and that, when p53 was genetically ablated, this resulted in too many sympathetic neurons (Slack et al. 1996; Aloyz et al. 1998). When p73 was discovered several years later (Kaghad et al. 1997; Jost et al. 1997), we examined the p73−/− mice (Yang et al. 2000) and made the unexpected finding that ΔNp73, the predominant isoform expressed in the developing nervous system, was essential for the survival of developing sympathetic neurons (Pozniak et al. 2000). Subsequent work showed that ΔNp73 promoted survival in part by antagonizing the proapoptotic actions of p53 but that it also acted by an uncharacterized p53-independent mechanism (Lee et al. 2004). The nature of this other mechanism was clarified upon examination of the third family member, p63 (Jacobs et al. 2005). This study showed that developing sympathetic neurons expressed only full-length TAp63 and that TAp63 was sufficient to cause sympathetic neuron apoptosis. Importantly, p63−/− sympathetic neurons did not die appropriately either in vivo or in culture, indicating that TAp63 functioned as an essential proapoptotic protein in these neurons. Together, these studies led to a model wherein the ultimate life versus death of a developing neuron was determined by the relative levels of the full-length proapoptotic family members TAp63 and p53 versus the dominant-inhibitory, prosurvival family member ΔNp73 (Jacobs et al. 2006; Fig. 2).

These findings suggested that ΔNp73 might be a general neuronal prosurvival protein. The first indication that this was indeed the case came from the p73−/−

Fig. 2 Survival of neurons and neural precursors is determined by the relative levels of full-length, proapoptotic versus ΔN truncated, prosurvival members of the p53 family. Shown are the family members that have been demonstrated to play an endogenous survival or death role in the mammalian nervous system

mice; these mice displayed greatly enlarged ventricles, reduced cortical tissue, and hippocampal dysgenesis (Yang et al. 2000). A thorough analysis of these CNS phenotypes demonstrated that, in the absence of p73, postnatal cortical neurons were gradually lost due to apoptosis and that this ultimately caused cortical thinning and ventricular enlargement (Pozniak et al. 2002). Since ΔNp73 is the predominant CNS isoform, these findings supported the idea that ΔNp73 was a general prosurvival protein and suggested that it was required for long-term neuronal maintenance. Additional support for these conclusions came from a study showing that adult p73+/− neurons were more vulnerable to insults such as DNA damaging agents or axonal injury (Walsh et al. 2004). The final proof that ΔNp73 was the relevant prosurvival isoform came from analysis of mice specifically lacking TAp73 or ΔNp73. Mice lacking TAp73 did not show neuronal apoptosis or enlarged ventricles, although they did display hippocampal dysgenesis (discussed below; Tomasini et al. 2008). In contrast, mice lacking ΔNp73 had normal hippocampi but exhibited deficits in neuronal survival (Wilhelm et al. 2010; Tissir et al. 2009). Intriguingly, the extent of neuronal loss in the ΔNp73-specific knockouts was not as profound as that seen in p73−/− mice, indicating that an as-yet-unknown interplay between TAp73 and ΔNp73 was required for the full manifestation of this phenotype. Nonetheless, these studies conclusively demonstrated that ΔNp73 is a potent and necessary survival protein that functions to enhance neuronal survival during developmental neuron death and to promote the long-term maintenance of mature neurons.

3 p73 Regulates Neurodegeneration

The finding that p73 was essential for long-term maintenance of CNS neurons raised the possibility that it might play some role in neuronal vulnerability in the face of aging and/or neurodegenerative disorders. To address this possibility, we examined the p73+/− mice (Wetzel et al. 2008), which displayed no overt phenotypes. Behavioral analysis of these mice demonstrated that, as they reached the age of 16–18 months, they developed a clasping behavior that had previously been seen in many mouse models of neurodegeneration. Moreover, they displayed age-dependent decline on a number of other behavioral/cognitive measurements, including the hang test, the open field test, and the Morris water maze. These behavioral perturbations were accompanied by neuroanatomical changes typical of the degenerating mammalian brain, including neuronal loss and neuronal degeneration as indicated by silver staining, inflammation, and lipofuscin accumulation. Perhaps most surprisingly, p73+/− mice developed aberrant neural phospho–tau-positive aggregates that were immunostained by antibodies that recognize phospho–tau in AD brain paired helical filaments. Thus, haploinsufficiency for p73 led to age-dependent neurodegeneration and development of neural phospho–tau aggregates.

If p73 haploinsufficiency causes neurodegeneration as animals age, does it also increase vulnerability to neurodegenerative disorders? Since phospho–tau-positive paired helical filaments are one of the hallmarks of AD, we chose to ask this question by crossing p73+/− mice to the TgCRND8 AD mouse model. These mice encode a double-mutant form of APP 695 (KM670/671NL+V717F) under control of the PrP gene promoter (Chishti et al. 2001). TgCRND8 mice develop plaques by two to three months of age and phospho–tau-positive aggregates by 9–10 months (Bellucci et al. 2007) but do not display neuronal loss. In contrast, the p73+/−, TgCRND8 crossed mice displayed neuronal loss and phospho–tau-positive aggregates as early as 1.5–2 months of age. Thus, haploinsufficiency for p73 sensitized mice to the pathological sequelae of mutant APP overexpression, with the result that the crossed animals had many of the phenotypes seen in human AD. Whether or not p73 haploinsufficiency generally sensitizes animals to neurodegenerative insults is a question we are currently asking.

These findings raised a number of key questions. First, how does p73 normally function to inhibit neurodegeneration? We believe that the neuronal loss and degeneration seen in p73+/− mice can be attributed to a decrease in ΔNp73-mediated antagonism of p53. In this regard, excitotoxicity has been implicated in many neurodegenerative disorders (Lin and Beal 2006), and p53 is necessary for excitotoxicity-induced neuronal death (reviewed in Jacobs et al. 2006). Moreover, the loss of p73 sensitizes neurons not just to apoptotic death but also to excitotoxicity-induced necrosis (Wetzel et al. 2008). However, while antagonism between ΔNp73 and p53 might explain the neuronal death and degeneration phenotypes, it is unlikely to explain the phospho–tau aggregates observed in p73+/− mice. Instead, we propose that this latter phenotype is due to ΔNp73-mediated regulation of c-jun N-terminal kinase (JNK), since JNK directly phosphorylates tau (Reynolds et al. 1997) and is important for excitotoxicity-induced neuronal death (Yang et al. 1997). In support of this idea, when ΔNp73 was ectopically expressed, it interacted with and inhibited JNK (Lee et al. 2004). Moreover, cultured p73+/− and p73−/− cortical neurons showed enhanced JNK activity, and the phospho–tau-positive neurons seen in p73+/− mice were also positive for the phosphorylated, activated form of JNK (Wetzel et al. 2008). Thus, we propose that ΔNp73 has multiple targets that together function to maintain the longevity and integrity of mammalian neurons and, when its expression is decreased, aberrant activation of these targets causes neurodegeneration.

A second key question coming from these findings is whether decreased ΔNp73 actually occurs within the human population and, if so, whether it enhances susceptibility to neurodegenerative disorders. Decreased ΔNp73 levels might occur via a number of different mechanisms. First, TAp73 and ΔNp73 are transcribed from two different promoters, and it is thus possible that selective transcriptional downregulation of ΔNp73 might occur. Such a mechanism has been postulated in a small study describing an association between p73 and AD in humans (Li et al. 2004). Second, some humans may be haploinsufficient for p73, since copy number variations have been reported in and around the p73 locus in humans (Wong et al. 2007; Redon et al. 2006). Third, there are polymorphisms in the p73 gene (for example, see Scacchi et al. 2009), and they could possibly decrease ΔNp73 levels through a variety of transcriptional and/or post-transcriptional mechanisms. Fourth, neuronal ΔNp73 levels are directly regulated by neurotrophic factors (Pozniak et al. 2000), and deficits in growth factors have long been thought to play a role in neurodegenerative disorders. Finally, the stability of ΔNp73 is a major regulatory point for this protein and it could potentially be modified. In this regard, genetic deletion of Pin1, a prolyl isomerase that stabilizes p73, also causes neurodegeneration and accumulation of phospho–tau aggregates (Liou et al. 2003), suggesting that perhaps p73 is a relevant Pin1 target within this context. There are, therefore, multiple ways to dysregulate ΔNp73, and whether these disruptions actually occur in humans and predispose them to neurodegeneration are important questions for the future. These findings also suggest a novel therapeutic strategy for neurodegenerative disorders based upon finding small molecules that can take advantage of these regulatory mechanisms to increase levels of ΔNp73 and thus increase neuronal life span.

4 The p53 Family Regulates Aging in Part by Regulating Adult Stem Cell Pools

While the aforementioned studies implicate p73 in nervous system aging, a number of other studies have implicated p53 and p63 in aging outside the nervous system. In particular, a number of mouse models with enhanced p53 activity display accelerated aging and decreased life span (Tyner et al. 2002; Lavigueur et al. 1989; Maier et al. 2004). Conversely, aging is enhanced in p63+/− mice or in mice with an inducible knockout of p63 in epithelial cells (Keyes et al. 2005). These phenotypes are thought to occur as a consequence of disruptions in p53 family-mediated regulation of genomic stability and cellular senescence, which in turn impact on cellular/organismal aging (Papazoglu and Mills, 2007). Intriguingly, a number of recent studies indicate that the p53 family might prevent aging by maintaining stem cell pools (Senoo et al. 2007; Suh et al. 2006; Su et al. 2009). Of particular relevance, a recent report showed that TAp63 prevented premature skin aging by promoting maintenance of adult skin stem cells (Su et al. 2009), supporting the emerging notion that tissues age because their tissue stem cells are depleted (Sharpless and DePinho 2007). This concept has recently been extended to nervous system aging and degeneration (Steiner et al. 2006; Galvan and Bredesen 2007), since we now know that the adult CNS contains neural stem cells that function to generate glial cells and certain neuronal subtypes on an ongoing basis, and a number of studies indicate that they also serve as a reservoir of stem cell activity following neural injury (reviewed in Zhao et al. 2008; Miller and Gauthier-Fisher 2009). Thus, adult NSCs might act to replace dead or degenerating cells during aging and/or in neurodegenerative disorders, and depletion of adult NSCs might therefore enhance neural vulnerability. While this is a new idea with little direct supporting evidence as of yet, a number of reports indicate that the p53 family regulates neural precursor survival, and we propose here that this might be one way that this family prevents premature neural aging.

5   The p53 Family and Neural Precursor Cells

While we have long known that the brain is "built" from embryonic neural precursors (the term precursors refers to both neural stem cells and more restricted progenitors), we have only recently appreciated the importance of NSCs in the adult brain. In particular, adult NSCs persist in two well-defined brain regions, the subependymal region of the lateral ventricles (SVZ) and the subgranular zone of the hippocampal dentate gyrus (SGZ; reviewed in Zhao et al. 2008; Miller and Gauthier-Fisher 2009). The adult NSCs found in the SVZ are a subset of forebrain lateral ventricle embryonic precursors that are maintained into adulthood and continuously generate interneurons that migrate via the rostral migratory stream (RMS) to the olfactory bulb and glial cells such as oligodendrocytes. Development of the hippocampal SGZ NSCs is somewhat different. During embryogenesis, a pool of hippocampal precursors generates the primary pyramidal neurons and a population of precursors that migrate away from the lateral ventricles. These latter precursors generate the dentate gyrus granule cells during the first postnatal week. Some of them also persist as adult SGZ NSCs and continue to generate granule cells and astrocytes.

If adult NSCs are indeed important for maintaining cognitive integrity and preventing neurodegeneration, it becomes critical to understand the endogenous mechanisms that regulate their numbers and longevity. Over the past several years, we and others have begun to ask if and how the p53 family regulates these different precursor populations. The first hint they might do so came from studies showing that some p53−/− embryos developed exencephaly (Sah et al. 1995; Armstrong et al. 1995). This phenotype is likely due to a deficit in p53-dependent neural precursor apoptosis resulting in an overgrowth of neural tissue, since we recently showed that p53 knockdown enhances survival of embryonic cortical precursors (Dugani et al. 2009), and Meletis et al. (2005) showed that adult p53−/− SVZ NSCs display enhanced survival. We then showed that ΔNp63 was expressed in and essential for the survival of precursors and newly born neurons in the embryonic cortex, a role it fulfilled by antagonizing p53 (Dugani et al. 2009). Intriguingly, ΔNp63 is also expressed in postnatal NSCs (unpublished data), suggesting that it might play a similar prosurvival role for adult NSCs. Thus, enhanced p53 activity due to reductions in ΔNp63 levels might perturb survival of neural precursors, leading to aberrant establishment or maintenance of the adult NSC pools needed to last for the lifetime of the animal and thereby potentially causing premature neural aging or vulnerability to neurodegenerative insults.

While decreased survival would be one way to deplete adult NSCs, a second way would be to decrease their self-renewal. In this regard, our recent data indicate that while ΔNp73 regulates neuronal longevity, TAp73 might instead regulate neural precursor maintenance. The first clue that this might be the case came from analysis of isoform-specific p73 knockout mice, which showed that the hippocampal dysgenesis and neuronal survival phenotypes were separable; decreased neuronal survival was due to loss of ΔNp73 (Wilhelm et al. 2010; Tissir et al. 2009), whereas hippocampal dysgenesis was due to loss of TAp73 (Tomasini et al. 2008). Intriguingly, the TAp73−/− mice displayed selective loss of the dorsal lower blade of the hippocampal dentate gyrus, a region that is comprised of the last-born dentate gyrus neurons (Li and Pleasure 2007). This unique deficit has not been reported in other mouse mutants, but a very similar phenotype was observed when the hippocampus was irradiated shortly after birth to destroy the postnatal hippocampal NSCs (Czurko et al. 1997). This similarity suggested that the TAp73−/− hippocampal phenotype might be caused by depletion of postnatal hippocampal neural precursors. We have recently tested this idea and have found that TAp73 is essential for the maintenance of developing and adult neural precursor pools and that, in its absence, adult NSC pools and adult neurogenesis are depleted (Fujitani et al. 2010). These somewhat surprising findings therefore raise a number of key questions with regard to neurodegeneration. First, if p73 is important for maintenance of both neurons and adult NSC pools, what is the relative importance of deficits in these two different cellular activities for the neurodegeneration and cognitive dysfunction observed in the p73+/− mice? Second, is the maintenance of adult NSCs important for healthy neural aging and, if so, is depletion of adult NSC pools in and of itself sufficient to cause neurodegeneration? And finally, do genes that cause neurodegeneration do so, at least in part, by depleting functional adult NSC pools, as recently suggested (reviewed in Marlatt and Lucassen 2010)? While we don't yet know the answers to any of these questions, the suggestion that adult NSCs may play a role in neurodegenerative disorders certainly provides a new perspective on an old problem.

6 A Model for the p53 Family and Neurodegeneration

The evidence summarized above indicates that the p53 family subserves two essential functions in the nervous system and that dysregulation of these functions causes premature nervous system aging and neurodegeneration. First, ΔNp73 acts to promote survival and longevity of mammalian neurons by antagonizing p53 and, at least in the peripheral nervous system, full-length proapoptotic TAp63. This prosurvival activity is essential during naturally occurring neuronal death, during normal aging, and in the face of neural injury or insults such as excitotoxicity or mutant APP. Evidence indicates that ΔNp73 also inhibits JNK activation, thereby inhibiting age- or insult-dependent formation of phospho–tau aggregates and likely also inhibiting JNK-dependent neuronal death. Second, the p53 family serves to regulate the establishment and maintenance of adult NSC pools, ensuring that they persist for the lifetime of the animal. It likely does this in two ways. In the embryo,

ΔNp63 functions to promote the survival of developing neural precursors and newly born neurons, likely by inhibiting p53. We propose that it plays a similar role in the adult nervous system, promoting survival of adult NSCs and their newly born neuronal progeny. In addition, we propose that TAp73 acts in neural precursors to promote their maintenance and self-renewal and that, in its absence, NSC pools become depleted and neurogenesis decreased. This process would be directly analogous to adult skin, where TAp63 functions to promote adult skin stem cell maintenance and to thereby prevent premature aging and aberrant wound-healing (Su et al. 2009). We propose that perturbation of either of these two activities, maintenance of neuronal longevity or adult NSC pools, would ultimately contribute to neurodegeneration and that, when both of these activities are coincidently perturbed, as is seen in mice haploinsufficient for the p73 gene, this would be sufficient on its own to cause age-dependent neurodegeneration. Whether or not these mechanisms are as important in humans as they are in mice is a key question for the future.

References

Aloyz RS, Bamji SX, Pozniak CD, Toma JG, Atwal J, Kaplan DR, Miller FD (1998) p53 is essential for developmental neuron death as regulated by the TrkA and p75 neurotrophin receptors. J Cell Biol 143:1691–1703

Armstrong JF, Kaufman MH, Harrison DJ, Clarke AR (1995) High-frequency developmental abnormalities in p53-deficient mice. Curr Biol 5:931–936

Bellucci A, Rosi MC, Grossi C, Fiorentini A, Luccarini I, Casamenti F (2007) Abnormal processing of tau in the brain of aged TgCRND8 mice. Neurobiol Dis 27:328–338

Chishti MA, Yang DS, Janus C, Phinney AL, Home P, Pearson J, Strome R, Zuker N, Loukides J, French J, Turner S, Lozza G, Grilli M, Kunicki S, Morrisette C, Paquette J, Gervais F, Bergeron C, Fraser PE, Carlson GA, George-Hyslop PS, Westaway D (2001) Early-onset amyloid deposition and cognitive deficits in transgenic mice expressing a double mutant form of amyloid precursor protein 695. J Biol Chem 276:21562–21570

Czurko A, Czeh B, Seress L, Nadel L, Bures J (1997) Severe spatial navigation deficit in the Morris water maze after single high dose of neonatal x-ray irradiation in the rat. Proc Natl Acad Sci USA 94:2766–2771

Dugani CB, Paquin A, Fujitani M, Kaplan DR, Miller FD (2009) P63 antagonizes p53 to promote the survival of embryonic neural precursor cells. J Neurosci 29:6710–6721

Dusart I, Guenet JL, Sotelo C (2006) Purkinje cell death: differences between developmental cell death and neurodegenerative death in mutant mice. Cerebellum 5:163–173

Fujitani M, Cancino GI, Dugani CB, Weaver ICG, Gauthier-Fisher A, Paquin A, Mak TW, Wojtowicz MJ, Miller FD, Kaplan DR (2010) TAp73 acts via the bHLH Hey2 to promote long-term maintenance of neural precursors. Curr Biol (in press)

Galvan V, Bredesen DE (2007) Neurogenesis in the adult brain: implications for Alzheimer's disease. CNS Neurol Disord Drug Targets 6:303–310

Herrup K, Yang Y (2007) Cell cycle regulation in the postmitotic neuron: oxymoron or new biology? Nat Rev Neurosci 8:368–378

Jacobs WB, Govoni G, Ho D, Atwal JK, Barnabé-Heider F, Keyes WM, Mills AA, Miller FD, Kaplan DR (2005) P63 is an essential proapoptotic protein during neural development. Neuron 48:743–756

Jacobs WB, Kaplan DR, Miller FD (2006) The p53 family in nervous system development and disease. J Neurochem 97:1571–1584

Jost CA, Marin MC, Kaelin WG Jr (1997) p73 is a simian [correction of human] p53-related protein that can induce apoptosis. Nature 389:191–194

Kaghad M, Bonnet H, Yang A, Creancier L, Biscan JC, Valent A et al (1997) Monoallelically expressed gene related to p53 at lp36, a region frequently deleted in neuroblastoma and other human cancers. Cell 90:809–819

Kaplan DR, Miller FD (2000) Neurotrophin signaling mechanisms. Curr Opin Neurobiol 10:381–391

Keyes WM, Wu Y, Vogel H, Guo X, Lowe SW, Mills AA (2005) P63 deficiency activates a program of cellular senescence and leads to accelerated aging. Genes Dev 19:1986–1999

Lavigueur A, Maltby V, Mock D, Rossant J, Pawson T, Bernstein A (1989) High incidence of lung, bone and lymphois tumours in transgenic mice overexpressing mutant alleles of the p53 oncogene. Mol Cell Biol 9:3982–3991

Lee AF, Ho DK, Walsh GS, Kaplan DR, Miller FD (2004) Evidence for a ΔNp73:p53 survival checkpoint upstream of the mitochondrion. J Neurosci 24:9174–9184

Li G, Pleasure SJ (2007) Genetic regulation of dentate gyrus morphogenesis. Prog Brain Res 163:143–152

Li Q, Athan ES, Wei M, Yuan E, Rice SL, Vonsattel JP, Mayeux RP, Tycko B (2004) TP73 allelic expression in human brain and allele frequencies in Alzheimer's disease. BMC Med Genet 5:14

Lin MT, Beal MF (2006) Mitochondrial dysfunction and oxidative stress in neurodegenerative disease. Nature 443:787–795

Liou YC, Sun A, Ryo A, Zhou XZ, Yu ZX, Huang HK, Uchid T, Bronson R, Bing G, Li X, Hunter T, Lu KP (2003) Role of the prolyl isomerase Pin1 in protecting against age-dependent neurodegeneration. Nature 424:556–561

Maier B, Gluba W, Bernier B, Turner T, Mohammad K, Guise T, Sutherland A, Thorner M, Scrable H (2004) Modulation of mammalian life span by the short isoform of p53. Genes Dev 18:306–319

Marlatt MW, Lucassen PJ (2010) Neurogenesis and Alzheimer's disease: biology and pathophysiology in mice and men. Curr Alzheimer Res 7:113–125

Meletis K, Wirta V, Hede SM, Nister M, Lundeberg J, Frisen J (2005) P53 suppresses the self-renewal of adult neural stem cells. Development 133:353–369

Miller FD, Gauthier-Fisher A (2009) Home at last: neural stem cell niches defined. Cell Stem Cell 4:507–510

Murray-Zmijewski F, Lane DP, Bourdon JC (2006) p53/p63/p73 isoforms: an orchestra of isoforms to harmonise cell differentiation and response to stress. Cell Death Differ 13:962–972

Papazoglu C, Mills AA (2007) P53: at the crossroad between cancer and ageing. J Pathol 211:124–133

Pozniak CD, Radinovic S, Yang A, McKeon F, Kaplan DR, Miller FD (2000) An anti-apoptotic role for the p53 family member, p73, during developmental neuron death. Science 289:304–6

Pozniak CD, Barnabe-Heider F, Rymar VV, Lee AF, Sadikot AF, Miller FD (2002) p73 is required for survival and maintenance of CNS neurons. J Neurosci 22:9800–9809

Redon R, Ishikawa S, Fitch KR, Feuk L, Perry GH, Andrews TD, Fiegler H, Shapero MH, Carson AR et al (2006) Global variation in copy number in the human genome. Nature 444:444–454

Reynolds CH, Utton MA, Gibb GM, Yates A, Anderton BH (1997) Stress-activated protein kinase/c-jun N-terminal kinase phosphorylates tau protein. J Neurochem 68:1736–1744

Sah VP, Attardi LD, Mulligan GJ, Williams BO, Bronson RT, Jacks T (1995) A subset of p53-deficient embryos exhibit exencephaly. Nat Genet 10:175–180

Scacchi R, Gambina G, Moretto G, Corbo RM (2009) Association study between p53 and p73 gene polymorphisms and the sporadic late-onset form of Alzheimer's disease. J Neural Transm 116:1179–1184

Senoo M, Pinto F, Crum CP, McKeon F (2007) p63 is essential for the proliferative potential of stem cells in stratified epithelia. Cell 129:523–536

Sharpless NE, DePinho RA (2007) How stem cells age and why this makes us grow old. Nat Rev Mol Cell Biol 8:703–713

Slack RS, Belliveau DJ, Rosenberg M, Atwal J, Lochmuller H, Aloyz R, Haghighi A, Lach B, Seth P, Cooper E, Miller FD (1996) Adenovirus-mediated gene transfer of the tumor suppressor, p53, induces apoptosis in postmitotic neurons. J Cell Biol 135:1085–1096

Steiner B, Wolf S, Kempermann G (2006) Adult neurogenesis and neurodegenerative disease. Regen Med 1:15–28

Su X, Paris M, Gi YJ, Cho MS, Lin Y, Lin L, Biernaskie JA, Sinha S, Prives C, Miller FD, Flores ER (2009) TAp63 prevents premature aging by promoting adult stem cell maintenance. Cell Stem Cell 5:64–75

Suh EK, Yang A, Kettenbach A, Bamberger C, Michaelis AH, Zhu Z, Elvin JA, Bronson RT, Crum CP, McKeon F (2006) p63 protects the female germ line during meiotic arrest. Nature 444:624–628

Tissir F, Ravni A, Achouri Y, Riethmacher D, Meyer G, Goffinet AM (2009) DeltaNp73 regulates neuronal survival in vivo. Proc Natl Acad Sci USA 106:16871–16876

Tomasini R, Tsuchihara K, Wilhelm M, Fujitani M, Rufini A, Cheung CC, Khan F, Itie-Youten A, Wakeham A, Tsao MS, Iovanna JL, Squire J, Jurisica I, Kaplan D, Melino G, Jurisicova A, Mak TW (2008) TAp73 knockout shows genomic instability with infertility and tumor suppressor functions. Genes Dev 22:2677–2691

Tyner SD, Venkatachalam S, Choi J, Jones S, Ghebranious N, Igelmann H, Lu X, Soron G, Cooper B, Brayton C, Hee Park S, Thompson T, Karsenty G, Bradley A, Donehower LA (2002) P53 mutant mice display early ageing-associated phenotypes. Nature 415:45–53

Vogelstein B, Lane D, Levine AJ (2000) Surfing the p53 network. Nature 408:307–310

Walsh GS, Orike N, Kaplan DR, Miller FD (2004) The invulnerability of adult neurons: a critical role for p73. J Neurosci 24:9638–9647

Wetzel MK, Naska S, Laliberte CL, Rymar VV, Fujitani M, Biernaskie JA, Cole CJ, Lerch JP, Spring S, Wang SH, Frankland PW, Henkelman RM, Josselyn SA, Sadikot AF, Miller FD, Kaplan DR (2008) P73 regulates neurodegeneration and phospho-tau accumulation in aging and Alzheimers disease. Neuron 59:708–721

Wilhelm MT, Rufini A, Wetzel MK, Tsuchihara K, Tomasini R, Itie-Youten A, Wakeham A, Arsenian-Henriksson M, Melino G, Kaplan DR, Miller FD, Mak TW (2010) Isoform-specific p73 knockout mice reveal a novel role for ΔNp73 in the DNA damage response pathway. Genes Dev 24:549–260

Wong KK, deLeeuw RJ, Dosanjh NS, Kimm LR, Cheng Z, Horsman DE, MacAulay C, Ng RT, Brown CJ, Eichler EE, Lam WL (2007) A comprehensive analysis of common copy-number variations in the human genome. Am J Hum Genet 80:91–104

Yang DD, Kuan CY, Whitmarsh AJ, Rincon M, Zheng TS, Davis RJ, Rakic P, Flavell RA (1997) Absence of excitotoxicity-induced apoptosis in the hippocampus of mice lacking the Jnk3 gene. Nature 389:865–870

Yang A, Kaghad M, Wang Y, Gillett E, Fleming MD, Dotsch V, Andrews NC, Caput D, McKeon F (1998) p63, a p53 homolog at 3q27-29, encodes multiple products with transactivating, death-inducing, and dominant-negative activities. Mol Cell 2:305–316

Yang A, Walker N, Bronson R, Kaghad M, Oosterwegel M, Bonnin J, Vagner C, Bonnet H, Dikkes P, Sharpe A, McKeon F, Caput D (2000) p73-deficient mice have neurological, pheromonal and inflammatory defects but lack spontaneous tumours. Nature 404:99–103

Zhao C, Deng W, Gage FH (2008) Mechanisms and functional implication of adult neurogenesis. Cell 132:645–660

p53, a Molecular Bridge Between Alzheimer's Disease Pathology and Cancers?

Frédéric Checler, Julie Dunys, Raphaëlle Pardossi-Piquard, and Cristine Alves da Costa

Abstract Cancers are characterized by enhanced cell survival and altered differentiation processes whereas Alzheimer's disease (AD)-affected brains exhibit exacerbated neuronal loss and cell death. Interestingly, several studies have consistently reported on an inverse relationship between cancer and AD. On the other hand, p53, a tumor-suppressor oncogene, is mutated and inactivated in a majority of human cancers; conversely, several lines of evidence concur to suggest an elevation of p53 and its transcriptional targets in AD brains. Therefore, one could envision p53 as a molecular bridge between cancer and AD pathologies. Although the role of p53 in cancer likely results from its inactivation by somatic mutations, the mechanistic aspects underlying a dysfunction in the control of p53 in AD had not been delineated. Here we survey recent evidence that p53 could control and be controlled by several members of the presenilin-dependent γ-secretase complex, and we briefly discuss the possibility that a functional deficit in presenilins could contribute to the genesis of a subset of tumors.

1 Introduction

Recent data indicated that people with a history of cancer display a lower probability of developing AD whereas, conversely, AD-affected patients have a slower rate of cancer development (Roe et al. 2009). Interestingly, this inverse relationship appeared to be specific to these two pathologies since cancer was not linked to vascular dementia (Roe et al. 2010). Intuitively, one could anticipate that a dysfunction in the control of the cell death process could underlie these observations. Thus, cancers are characterized by anarchical cell proliferation whereas AD pathology, as most of the neurodegenerative diseases, displays enhanced cell death.

F. Checler (✉)
IPMC and IN2M, UMR6097 CNRS/UNSA, Team Fondation pour la Recherche Médicale, Sophia-Antipolis, Valbonne 06560, France
e-mail: checler@ipmc.cnrs.fr

T. Curran and Y. Christen (eds.), *Two Faces of Evil: Cancer and Neurodegeneration*, Research and Perspectives in Alzheimer's Disease, DOI 10.1007/978-3-642-16602-0_8, © Springer-Verlag Berlin Heidelberg 2011

These considerations argue in favor of the existence of common molecular triggers that could be either down regulated or inactivated in cancer and likely exacerbated in AD.

Several lines of consideration and recent data identify p53 as a putative key protein for both pathologies. p53 is an oncogene, the transcription factor activity of which controls complex stress signals-induced cellular responses leading to cell-cycle arrest (Levine and Oren 2009). This "guardian of the genome" (Levine and Oren 2009) is very often mutated in cancers, indicating that the inactivation of its function could be directly linked to the development of spontaneous tumors. On the other hand, the expression of p53 is enhanced in AD-affected brains (Kitamura et al. 1997; de la Monte et al. 1998; Garcia-Ospina et al. 2003; Ohyagi et al. 2005) and is likely responsible for AD-linked neuronal cell death (Cotman and Anderson 1995; Shimohama 2000). While several p53 transcriptional targets have been identified and directly linked to the control of cell cycle (Levine and Oren 2009), the putative targets linking p53 to AD have not been delineated. Recent data from our laboratory focused on the putative role of p53 in the control of amyloid β-peptides (Aβ) production (Checler et al. 2010).

In AD-affected brains, Aβ peptides accumulate as extracellular deposits in senile plaques (Suh and Checler 2002). These peptides are generated by subsequent cleavages by proteolytic activities referred to as β- and γ-secretases (Checler 1995). The latter activity mainly corresponds to a high molecular weight complex that includes at least four proteins, namely presenilins (PS) 1 or 2, Aph-1, Pen-2 and nicastrin (Takasugi et al. 2003). Interestingly, PS deficiency leads to keratocarcinoma in mice (Tournoy et al. 2004), and skin lesion severity is directly linked to a gene dose-dependent mechanism (Xia et al. 2001; Tournoy et al. 2004). Here, we will survey data showing that PS1/2 could directly modulate p53 at a transcriptional level, and we will document a molecular feedback by which p53 controls certain members of the γ-secretase complex.

2 p53 and Members of the γ-Secretase Complex: an Intimate Functional Cross-Talk

AD is characterized by two main anatomical lesions called senile plaques and neurofibrillary tangles (Selkoe 1991). The former are due to the accumulation of a set of hydrophobic peptides referred to as amyloid β-peptides (Aβ). Aβ peptides derive from the proteolytic processing of a transmembrane protein (βAPP) that undergoes sequential cleavages by β- and γ-secretases (Checler 1995). BACE1 (β- site APP cleaving enzyme 1) is an acidic protease that accounts for β-secretase activity in brain (for review, see Vassar, 2001), whereas γ-secretase exists as PS-independent (Armogida et al. 2001; Wilson et al. 2003) and PS-dependent activities (De Strooper et al. 1998). The latter corresponds to a multi-proteic complex made of at least four distinct components, namely PS1 or PS2, Aph-1,

Pen-2 and nicastrin (Takasugi et al. 2003). It is thought that PS harbor the catalytic core of the γ-secretase complex, although several theoretical and mechanistic issues still preclude its consideration as a definite dogma (Checler 2001).

Mutations on βAPP and PS all lead to an alteration of Aβ peptides (Ancolio et al. 1999; Checler 1999). Most often, Aβ species generated by, or from, mutated proteins display distinct N-terminal moieties but systematically exhibit a dipeptide C-terminal extension that exacerbates their insolubility and renders these species prone to aggregation (Burdick et al. 1992). These biophysical modifications have functional consequences. Thus, Aβ42 is more toxic than Aβ40 and triggers apoptosis. Interestingly, Ohyagi and colleagues (2005) demonstrated that intracellular Aβ42 triggers its proapoptotic phenotype by activating the promoter transactivation of p53. Since previous studies demonstrated that PS1 and PS2 mutations enhance cell death (Guo et al. 1999; Alves da Costa et al. 2002, 2003) via a p53-dependent mechanism (Alves da Costa et al. 2002, 2003, 2006), one could envision that pathogenic mutations trigger Aβ-mediated and p53-dependent proapoptotic phenotypes. Recent data indicate that an additional βAPP catabolite generated by γ-secretase could also contribute to p53-dependent cell death. Thus, we recently established that AICD (βAPP intracellular domain), the C-terminal intracellular counterpart generated by γ-secretase, could harbor transcription factor properties (Pardossi-Piquard et al. 2005; Alves da Costa et al. 2006) and could enhance p53 promoter transactivation (Alves da Costa et al. 2006). Therefore, PS mutations increase the production of two βAPP catabolites that are able to significantly enhance p53 cellular pathway.

Although mutations on PS1 and PS2 all exacerbate p53-dependent cell death, it is striking that wild-type PS1 and PS2 display distinct abilities to control cellular vulnerability. Thus, PS1 lowers neuronal susceptibility to apoptosis (Bursztajn et al. 1998) whereas PS2 displays an obvious pro-apoptotic phenotype (Wolozin et al. 1996; Araki et al. 2000; Alves da Costa et al. 2002), which was again clearly linked to p53 since wild-type PS1 lowers p53 whereas, conversely, PS2 enhances this pathway (Alves da Costa et al. 2006). This finding was at first sight puzzling if one considers that both PS display the structural catalytic ability to generate AICD. Therefore, if AICD is triggering p53 promoter activation, how could PS1 lower p53-dependent cell death? This set of data has been explained by the interplay between the two PS, with each of them being able to control the expression of its congener (Alves da Costa et al. 2002, 2006; Kang et al. 2005). Overall, this functional dialogue between the two family members allows the control of AICD homeostasis and, thereby, the level of p53. Since full depletion of both PS1 and PS2 yields an anti-apoptotic phenotype similar to that exhibited by PS2 invalidation only (Alves da Costa et al. 2006), the corollary of these considerations is that PS2 is clearly dominant over PS1 concerning the control of p53-dependent cell death.

p53-dependent apoptosis is also controlled by other members of the γ-secretase complex. Thus, we established that Aph-1a and Pen-2 both reduced p53-dependent staurosporine-induced caspase-3 activation in several cell systems from neuronal and human origins (Dunys et al. 2007). More recently, we showed that nicastrin displayed a similar protective phenotype (Pardossi-Piquard et al. 2009). These studies were in good agreement with other data indicating that Aph-1a, Pen-2 or nicastrin invalidation in mice or zebrafish led to severe apoptosis in brain and heart of embryos (Serneels et al. 2005; Campbell et al. 2006; Van Nguyen et al. 2006). The important conclusions of these studies concerned the degree of participation of the γ-secretase complex structure and/or activity in the observed functions displayed by Aph-1a, Pen-2 and nicastrin. Thus, Aph-1a and Pen-2-associated p53-dependent protective functions require the γ-secretase structure but are independent of γ-secretase activity. On the other hand, nicastrin apparently does not require the γ-secretase activity. Therefore, members of the γ-secretase could modulate p53-dependent cell death inside or outside the γ-secretase complex per se, which clearly adds another level of complexity to the complex!

Interestingly, a feedback loop exists by which p53 modulates the promoter transactivation of several members of the γ-secretase complex. Thus it has been documented that p53 acts as a transcriptional repressor of PS1 promoter (Roperch et al. 1998; Pastorcic and Das 2000). Recently, we established that p53 deficiency triggered a decrease in Pen-2 expression, promoter transactivation and mRNA levels (Dunys et al. 2009), which was due to an indirect mechanism mediated by AICD (Dunys et al. 2009). This set of data indicates first that a feedback control of Pen-2 and p53 expressions exists. Thus, as described above, Pen-2 reduces p53 and, therefore, triggers a reduction of Pen-2 and thereby contributes to restore p53 levels. The other interesting aspect is the demonstration that PS could control Pen-2 by AICD-mediated p53-dependent process. Therefore, a member of the γ-secretase complex could regulate the expression of another one, which unravels another level of complexity for the control of γ-secretase expression/activity.

3 Cancer and AD: Is There a Molecular Link?

There are epidemiological clues of an inverse relationship between cancer and neurodegenerative pathologies. Interestingly, recent studies indicated that people developing AD have a lower rate of cancer history. Conversely, those affected by cancer display slower rates of senile dementia of the Alzheimer's type whereas those with vascular dementia remained unaffected (Roe et al. 2010). p53 is a natural suspect, the function of which could be exacerbated or inactivated in neurodegenerative diseases or cancers, respectively. Thus, p53 is inactivated by mutations in most human cancers. It is interesting that the relationship between AD and cancer cases was particularly striking for skin cancers (Behrens et al. 2009) and that PS deficiency drastically reduces p53 and triggers skin cancers (Xia et al. 2001; Tournoy et al. 2004). This reduction of p53 also triggered by proteins involved in other neurodegenerative disorders, such as Parkinson's disease (da Costa et al. 2009), highlights the fact that an inverse relationship between cancer and neurodegenerative diseases where p53 would play a key role could a more general feature.

Both Aβ42 and AICD could control p53 at a transcriptional level. In genetic cases of AD, where γ-secretase activity is enhanced, it is conceivable that increased

Aβ and AICD productions may lead to enhanced cell death and neurodegeneration. Obviously, those few cases that are aggressive and with early onset preclude any examination of associated rates of cancer. In the more numerous cases of sporadic origin, there are no clues of enhanced γ-secretase activity, and it is more likely that Aβ and AICD levels are enhanced through an age-related reduction of their inactivating proteases, namely neprilysin and insulin-degrading enzyme (Carson and Turner 2002; Pardossi-Piquard et al. 2005). Therefore, one can speculate that Aβ and AICD-associated increase of p53 could account for both exacerbated cell death and, therefore, explain the protection of those affected cases against tumogenicity. Overall, these findings identify p53 as a molecular link between AD and cancer pathologies.

References

Alves da Costa C, Paitel E, Mattson MP, Amson R, Telerman A, Ancolio K, Checler F (2002) Wild-type and mutated presenilins 2 trigger p53-dependent apoptosis and down-regulate presenilin 1 expression in HEK293 human cells and in murine neurons. Proc Natl Acad Sci USA 99:4043–4048

Alves da Costa C, Mattson MP, Ancolio K, Checler F (2003) The C-terminal fragment of presenilin 2 triggers p53-mediated staurosporine-induced apoptosis, a function independent of the presenilinase-derived N-terminal counterpart. J Biol Chem 278:12064–12069

Alves da Costa C, Sunyach C, Pardossi-Piquard R, Sevalle J, Vincent B, Boyer N, Kawarai T, Girardot N, St George-Hyslop P, Checler F (2006) Presenilin-dependent gamma-secretase-mediated control of p53-associated cell death in Alzheimer's disease. J Neurosci 26:6377–6385

Ancolio K, Dumanchin C, Barelli H, Warter JM, Brice A, Campion D, Frébourg T, Checler F (1999) Unusual phenotypic alteration of β amyloid precursor protein (βAPP) maturation by a new Val->Met βAPP-770 mutation responsible for probable early-onset Alzheimer's disease. Proc Natl Acad Sci USA 96:4119–4124

Araki W, Yuasa K, Takeda S, Shirotani K, Takahashi K, Tabira T (2000) Overexpression of presenilin-2 enhances apoptotic death of cultured cortical neurons. Ann NY Acad Sci 920:241–244

Armogida M, Petit A, Vincent B, Scarzello S, da Costa CA, Checler F (2001) Endogenous beta-amyloid production in presenilin-deficient embryonic mouse fibroblasts. Nat Cell Biol 3:1030–1033

Behrens MI, Lendon C, Roe CM (2009) A common biological mechanism in cancer and Alzheimer's disease? Curr Alzheimer Res 6:196–204

Burdick D, Soreghan B, Kwon M, Kosmoski J, Knauer M, Henschen A, Yates J, Cotman C, Glabe C (1992) Assembly and aggregation properties of synthetic Alzheimer's A4/β amyloid peptide analogs. J Biol Chem 267:546–554

Bursztajn S, DeSouza R, McPhie DL, Berman SA, Shioi J, Robakis NK, Neve RL (1998) Overexpression in neurons of human presenilin-1 or a presenilin-1 familial Alzheimer's disease mutant does not enhance apoptosis. J Neurosci 18:9790–9799

Campbell WA, Yang H, Zetterberg H, Baulac S, Sears JA, Liu T, Wong STC, Zhong TP, Xia W (2006) Zebrafish lacking Alzheimer presenilin enhancer 2 (Pen-2) demonstrate excessive p53-dependent apoptosis and neuronal loss. J Neurochem 96:1423–1440

Carson JA, Turner AJ (2002) β-amyloid catabolism: roles for neprilysin (NEP) and other metallopeptidases. J Neurochem 81:1–8

Checler F (1995) Processing of the β-amyloid precursor protein and its regulation in Alzheimer's disease. J Neurochem 65:1431–1444

Checler F (1999) Presenilins: multifunctional proteins involved in Alzheimer's disease pathology. Iubmb LIFE 48:33–39

Checler F (2001) The multiple paradoxes of presenilins. J Neurochem 76:1621–1627

Checler F, Dunys J, Pardossi-Piquard R, Alves da Costa C (2010) p53 is regulated by and regulates members of the gamma-secretase complex. Neurodegener Dis 7:50–55

Cotman CW, Anderson AJ (1995) A potential role for apoptosis in neurodegeneration and Alzheimer's disease. Mol Neurobiol 10:19–45

Da Costa CA, Sunyach C, Giaime E, West A, Corti O, Brice A, Safe S, Abou-Sleiman PM, Wood NW, Takahashi H, Goldberg MS, Shen J, Checler F (2009) Transcriptional repression of p53 by parkin and impairment by mutations associated with autosomal recessive juvenile Parkinson's disease. Nat Cell Biol 11:1370–1375

de la Monte S, Sohn YK, Ganju YK, Wands JR (1998) p53- and CD95-associated apoptosis in neurodegenerative diseases. Lab Invest 78:401–411

De Strooper B, Saftig P, Craessaerts K, Vanderstichele H, Guhde G, Annaert W, Von Figura K, Van Leuven F (1998) Deficiency of presenilin-1 inhibits the normal cleavage of amyloid precursor protein. Nature 391:387–390

Dunys J, Kawarai T, Sevalle J, Dolcini V, St George-Hyslop P, Alves da Costa C, Checler F (2007) p53-dependent Aph-1 and Pen-2 anti-apoptotic phenotype requires the integrity of the gamma-secretase complex but is independent of its activity. J Biol Chem 282:10516–10525

Dunys J, Sevalle J, Giaime E, Pardossi-Piquard R, Vitek MP, Renbaum P, Levy-Lahad E, Zhang YW, Xu H, Checler F, da Costa CA (2009) p53-dependent control of transactivation of the Pen2 promoter by presenilins. J Cell Sci 122:4003–4008

Garcia-Ospina GP, Ginenez-del Rio M, Lopera F, Velez-Pardo C (2003) Neuronal DNA damage correlates with a positive detection of c-Jun, nuclear factor κB, p53 and Par-4 transcription factors in Alzheimer's disease. Rev Neurol 36:1004–1010

Guo Q, Fu W, Sopher BL, Miller MW, Ware CB, Martin GM, Mattson MP (1999) Increased vulnerability of hippocampal neurons to excitotoxic necrosis in presenilin-1 mutant knock-in mice. Nat Med 5:101–106

Kang DE, Yoon IS, Repetto E, Busse T, Yermian N, Ie L, Koo EH (2005) Presenilins mediate PI3K/AKT and ERK activation via select signaling receptors: selectivity of PS2 in PDGF signaling. J Biol Chem 280:31537–31547

Kitamura Y, Shimohama S, Kamoshima W, Matsuoka Y, Nomura Y, Taniguchi T (1997) Changes of p53 in the brains patients with Alzheimer disease. Biochem Biophys Res Comm 232:418–421

Levine AJ, Oren M (2009) The first 30 years of p53: growing ever more complex. Nat Rev Cancer 9:749–758

Ohyagi Y, Asahara H, Chui DH, Tsuruta Y, Sakae N, Miyoshi K, Yamada T, Kikuchi H, Taniwaki T, Murai H, Ikezoe K, Furuya H, Kawarabayashi T, Shoji M, Checler F, Iwaki T, Makifuchi T, Takeda K, Kira I, Tabira T (2005) Intracellular Abeta42 activates p53 promoter: a pathway to neurodegeneration in Alzheimer's disease. FASEB J 19:255–257

Pardossi-Piquard R, Petit A, Kawarai T, Sunyach C, Alves da Costa C, Vincent B, Ring S, D'Adamio L, Shen J, Muller U, St George Hyslop P, Checler F (2005) Presenilin-dependent transcriptional control of the Abeta-degrading enzyme neprilysin by intracellular domains of betaAPP and APLP. Neuron 46:541–554

Pardossi-Piquard R, Dunys J, Giaime E, Guillot-Sestier MV, St George-Hyslop P, Checler F, Alves da Costa C (2009) p53-dependent control of cell death by nicastrin: lack of requirement for presenilin-dependent gamma-secretase complex. J Neurochem 109:225–237

Pastorcic M, Das HK (2000) Regulation of transcription of the human presenilin-1 gene by Ets transcription factor and the p53 protooncogene. J Biol Chem 275:34938–34945

Roe CM, Behrens MI, Xiong C, Miller JP, Morris JC (2009) Alzheimer disease and cancer. Neurology 64:895–898

Roe CM, Fitzpatrick AL, Xiong C, Sieh W, Kuller L, Miller JP, Williams MM, Kopan R, Behrens MI, Morris JC (2010) Cancer linked to Alzheimer disease but not vascular dementia. Neurology 74:106–112

Roperch J-P, Alvaro V, Prieur S, Tyunder M, Nemani M, Lethrosne F, Piouffre L, Gendron M-C, Israeli D, Dausset J, Oren M, Amson R, Telerman A (1998) Inhibition of presenilin1 expression is promoted by p53 and p21^{WAF-1} and results in apoptosis and tumor suppression. Nat Med 4:835–838

Selkoe DJ (1991) The molecular pathology of Alzheimer's disease. Neuron 6:487–498

Serneels L, Dejaegere T, Craessaerts K, Horré K, Jorissen E, Tousseyn T, Hébert S, Coolen M, Martens G, Zwijsen A, Annaert W, Hartmann D, De Strooper B (2005) Differential contribution of the three Aph1 genes to g-secretase activity in vivo. Proc Natl Acad Sci USA 102:1719–1724

Shimohama S (2000) Apoptosis in Alzheimers' disease- an update. Apoptosis 5:9–16

Suh YH, Checler F (2002) Amyloid precursor protein, presenilins, and alpha-synuclein: molecular pathogenesis and pharmacological applications in Alzheimer's disease. Pharmacol Rev 54:469–525

Takasugi N, Tomita T, Hayashi I, Tsuruoka M, Niimura M, Takahashi Y, Thinakaran G, Iwatsubo T (2003) The role of presenilin cofactors in the γ-secretase complex. Nature 422:438–441

Tournoy J, Bossuyt X, Snellinx A, Regent M, Garmyn M, Serneels L, Saftig P, Craessaerts K, De Strooper B, Hartmann D (2004) Partial loss of presenilins causes seborrheic keratosis and autoimmune disease in mice. Human Mol Gen 13:1321–1331

Van Nguyen V, Hawkins C, Bergeron C, Supala A, Huang J, Westaway D, St George-Hyslop P, Rozmahel R (2006) Loss of nicastrin elicits an apoptotic phenotype in mouse embryos. Brain Res 1086:76–84

Vassar R (2001) The β-secretase, BACE. J Mol Neurosci 17:157–170

Wilson CA, Doms RW, Lee VM-Y (2003) Distinct presenilin-dependent and presenilin-independent γ-secretases are responsible for total cellular Aβ production. J Neurosci Res 74:361–369

Wolozin B, Iwasaki K, Vito P, Ganjei JK, Lacana E, Sunderland T, Zhao B, Kusiak JW, Wasco W, D'Adamio L (1996) Participation of presenilin 2 in apoptosis: enhanced basal activity conferred by an Alzheimer mutation. Science 274:1710–1713

Xia X, Qian S, Soriano S, Wu Y, Fletcher AM, Wang XJ, Koo EH, Wu X, Zheng H (2001) Loss of presenilin 1 is associated with enhanced beta-catenin signaling and skin tumorigenesis. Proc Natl Acad Sci USA 98:10863–10868

RNA regulation in Neurodegeneration and Cancer

Robert B. Darnell

Abstract The paraneoplastic neurologic diseases are a set of brain degenerations that develop in the setting of occult cancer. These disorders arise when common cancers express brain proteins, triggering an anti-tumor immune response and tumor immunity. Studies of the nature of these brain-cancer proteins has revealed a new world of neuron-specific RNA binding proteins whose functions may be co-opted by tumor cells. The interest in their function has spawned the development of new methods and strategies to understand RNA-protein regulation in living tissues. These approaches, and the new biology they reveal, are discussed here.

1 Neurologic Disease and Cancer: The Paraneoplastic Neurologic Degenerations

The paraneoplastic neurologic degenerations (PNDs) are a set of brain disorders that develop in the setting of occult cancer (Table 1). These disorders arise when common cancers express brain proteins, triggering an anti-tumor immune response and, at least in some cases, clinically effective suppression of the malignancy. These disorders come to the attention of the patient and physician only when the tumor immune response breaches immune privilege of the nervous system and begins to attack the neurons that are normally expressing the tumor antigen (which we have termed onconeural antigens). Our laboratory was the first to demonstrate that genes encoding onconeural antigens can be cloned using high titer antisera from these patients (Darnell et al. 1989, 1991; McKeever and Darnell 1992; Newman et al. 1995; Darnell 1996), and currently over a dozen well-described PND antigens have been defined.

R.B. Darnell
Laboratory of Molecular Neuro-Oncology, Howard Hughes Medical Institute, The Rockefeller University, Box 226, 1230 York Ave, New York, NY 10021, USA
e-mail: darnelr@rockefeller.edu

T. Curran and Y. Christen (eds.), *Two Faces of Evil: Cancer and Neurodegeneration*,   103
Research and Perspectives in Alzheimer's Disease, DOI 10.1007/978-3-642-16602-0_9,
© Springer-Verlag Berlin Heidelberg 2011

Table 1 Representative List of Paraneoplastic Neurologic Disorders

Antibody	Subcellular location of antigen	Antigen/Gene(s)	Usual Tumor	Neurologic disorder
Anti-Hu	Nucleus and cytoplasm (all neurons)	HuD	SCLC, neuroblastoma, prostate	PEM, PSN, autonomic dysfunction
Anti-Yo	Cytoplasm, Purkinje cells	CDR2	Ovary, breast, lung	PCD
Anti-Ri	Nucleus and cytoplasm (CNS neurons)	Nova 1,2	Breast, Gyn, lung, bladder	Ataxia/opsoclonus
Anti-CRMP5 (CV2)	Cytoplasm oligodendrocytes, neurons	CRMP5	SCLC, thymoma	PEM, PCD, chorea, optic, sensory neuropathy
Anti-amphiphysin	Pre-synaptic	Amphiphysin	Breast, SCLC	SPS
Anti-Ma2	Neurons (nucleolus)	Ma-2	Testis	Limbic, brainstem encephalitis
Anti-NMDAR	Neurons, hippocampus	NR1/NR2	Ovarian teratoma	PEM
Anti-recoverin	Photoreceptor, ganglion cells	Recoverin	SCLC	CAR
Anti-AChR	Postsynaptic NMJ (electron immunohistochemistry)	AChR	Thymoma	MG
Anti-VGCC	Pre-synaptic NMJ	P/Q VGCC	SCLC	LEMS
Anti-GAD	Purkinje cell cytoplasm, nerve terminals, other neurons	Glutamic acid decarboxylase	Several (renal, Hodgkin, SCLC)	SPS
Anti-glycine receptor	Brain stem, spinal cord neurons	Glycine receptor	Lung cancer	PERM
Anti-GABA-BR	Neuronal surface, axons, dendrites	GABA-B receptor	SCLC	LE
Uncommon syndromes:				
Anti-gephyrin	Post-synaptic membranes	Gephyrin	Unknown primary	SPS
Anti-synaptotagmin	Presynaptic junction	Vesicle protein	??	LEMS
Anti-synaptophysin	Presynaptic junction	Vesicle protein	SCLC	Neuropathy
Anti-NB	Purkinje cell cytoplasm	Neuron specific vesicle coat	?? Ovary	PCD

Abbreviations: *SCLC* small-cell lung cancer; *PEM* paraneoplastic encephalomyelitis; *PSN* paraneoplastic sensory neuronopathy; *PCD* paraneoplastic cerebellar degeneration; *SPS* stiff person syndrome; *CNS* central nervous system; *Gyn* gynecological cancer; *CRMP* collapsing response-mediated protein 5; *CAR* cancer associated retinopathy; *NMDAR* N-N-methyl-D-aspaate receptor; *NMJ* neuromuscular junction; *VGCC* voltage-gated calcium channel

Cloning cDNAs encoding PND antigens have provided a wealth of information regarding the nature of the PND antigens (Musunuru and Darnell 2001; Dredge et al. 2001; Licatalosi and Darnell 2006) as well as their role in disease pathogenesis (Albert and Darnell 2004; Roberts and Darnell 2004; Darnell and Posner 2006). For example, the initial cloning of a cerebellar antigen from a woman with presumed ovarian cancer led to the discovery of a long-suspected, neuron-specific vesicle coat protein (Newman et al. 1995). Cloning of the cerebellar antigen (cdr2) that is associated more commonly with gynecologic cancer revealed the presence of antigen-specific T cells in these patients, establishing a cell-mediated component to the disorders as well as a means by which tumor antigen is likely cross-presented to dendritic cells to initiate the anti-tumor immune response (Albert et al. 1998, 2000). Subsequent cloning of the genes encoding T cell receptors from these patients offers a possible new therapeutic strategy, by passive transfer of these genes into T cells to induce tumor-cell targeting (Santomasso et al. 2007).

2   RNA Binding Proteins in Neurologic Disease and Cancer

Among the set of cloned PND antigens, perhaps none has generated as much biologic interest as the Nova and Elavl (Hu) family of neuron-specific RNA binding proteins. The Nova proteins (Buckanovich et al. 1993; Yang et al. 1998) are ectopically expressed in lung or gynecologic cancers and trigger neurologic symptoms (excess motor movements of the eyes and musculature) interpreted by neurologists as failure of inhibitory motor control. Studies of the Nova proteins have established what is arguably the best data available to understand a tissue-specific RNA regulatory protein in the brain; these findings will be discussed here. The Elavl proteins (originally termed the Hu proteins; Graus et al. 1985; Szabo et al. 1991) are still incompletely understood (Hinman and Lou 2008; Musunuru and Darnell 2001) but are believed to relate to post-transcriptional regulation of RNA in the brain. Many of the approaches used to understand the biology of Nova can be applied more generally to the understanding of the biology of the Elavl proteins and other RNA binding proteins in the brain and cancer.

3   Back to the Basics: RBP Functional Studies

Understanding RBP function requires understanding the protein's interaction with RNA. Nova has emerged as one of the best-understood mammalian RNA binding proteins in part because there is a good understanding of its RNA substrates. After recognizing that Nova harbors three KH-type RNA binding proteins (Buckanovich et al. 1996), in vitro RNA selection experiments (Green et al. 1991; Szostak and Ellington 1993) were applied to reveal (Buckanovich and Darnell 1997; Yang et al. 1998) that Nova recognizes a core 4 nucleotide repeat sequence $(UCAU)_3$. Mutagenesis and x-ray crystallographic studies identified the CA dinucleotide as a critical invariant component of Nova-RNA binding (Jensen et al. 2000; Lewis et al. 2000) with specificity restricted to bounding pyrimidines, hence the more general consensus of YCAY repeats as the core binding motif.

4 Genetic Systems and RBP Function

These studies prompted a search for brain transcripts harboring such repeat elements. This relatively pedestrian approach yielded three targets, which were characterized by mutagenesis, boundary mapping, and functional analyses in some detail. The first target transcript harboring a Nova-regulated YCAY cluster was the inhibitory glycine receptor $\alpha 2$ (GlyR$\alpha 2$). Interestingly, this cluster was located within an intronic sequence upstream of an alternatively spliced exon (E3A) of GlyR$\alpha 2$. Analysis of alternative splicing of minigenes transfected into tissue culture cells demonstrated that Nova acted specifically on the YCAY cluster to mediate an increase in inclusion of E3A (Buckanovich and Darnell 1997). Subsequently, analogous studies demonstrated that Nova was able to bind intronic YCAY elements to mediate an increase in the $\gamma 2L$ exon of the GABA$_A$ transcript (Dredge and Darnell 2003) and to bind exonic and intronic YCAY elements to autoinhibit splicing of Nova1 exon 4 (Dredge et al. 2005). This finding led to the hypothesis that Nova might regulate alternative splicing of these and other transcripts in neurons.

This hypothesis was tested in vivo by generating Nova null mice. GlyRa2 E3A and GABA$_A$ g2L splicing showed consistent two- and three-fold decreases in utilization, respectively, in Nova1 knockout (KO) mice (Jensen et al. 2000), whereas Nova E4 exclusion was decreased ~0.75-fold in Nova1 heterozygous mice (Dredge et al. 2005). These studies established Nova as the first bona-fide mammalian tissue-specific splicing factor whose actions were validated in live animals, setting the stage for more global studies of Nova function. These findings were further stimulated by recognizing the correlation between Nova targeting in PND patients, who show defects in inhibitory motor control, and the observation that 2/2 Nova-regulated transcripts encoded inhibitory neurotransmitter receptors (Darnell 2006).

5 Bioinformatics, Genetics and Biochemistry: Beginnings of a Holistic Approach to RBP Function

Initial studies aimed at discovering the general nature of Nova action on alternative splicing were done using a new (at the time) Affymetrix exon junction array. Applying wild type or Nova KO RNA to these arrays, whose probesets were designed to detect alternative splice junctions, revealed a wealth of new target transcripts that could be independently validated as Nova targets in vivo. Most surprisingly, the results revealed that the vast majority of Nova-regulated transcripts encoded synaptic proteins. More precisely, the steady-state level of these regulated RNAs was unaltered in Nova null mouse brain, but their quality –manifested by different ratios of alternative exons – was altered (Ule et al. 2005).

A limitation of this work came from careful parsing of the term "Nova-regulated transcripts." These transcripts showed Nova-dependent changes in alternative splicing, but such changes might be a direct action of Nova- for example, acting on YCAY elements – or an indirect action – for example, through regulation of a different splicing factor. Hence two series of studies were used to follow up the finding from the "correlative" Affymetrix exon junction array. A heuristic bioinformatic algorithm was developed to search for YCAY elements within this set of Nova targets, revealing an enrichment of elements within them (although the algorithm was limited, with ~50% false positive rate, in its ability to predict targets de novo; Ule et al. 2006).

Surprisingly, a position-dependent action of Nova on these transcripts was also discovered, such that Nova binding elements upstream or within alternative exons correlated with inhibition of exon inclusion whereas Nova binding downstream correlated with enhancement of exon inclusion (Ule et al. 2006). An effort to further strengthen these correlations was made, both by undertaking detailed in vitro splicing assays to provide a mechanistic basis for the positional effects and through analysis of splicing intermediates by RT-PCR, providing evidence for asymmetric actions on splicing intermediates that correlated with the position of the YCAY elements (Ule et al. 2006).

6   HITS-CLIP and the Development of a Comprehensive Approach to RBP Function

The heavy use of the word "correlation" in the previous section emphasizes what was missing from the microarray and bioinformatic studies: evidence of direct Nova-RNA interaction. Although such evidence was possible on a case-by-case basis, made most persuasively with Nova's action on the $GABA_A$ transcript (Dredge and Darnell 2003; Ule et al. 2006), it was not possible to undertake such studies, or develop an entirely compelling description of Nova action, on what began to approach 100 putative Nova-regulated transcripts. Hence a new approach was needed and was developed in the form of CLIP (cross-link-immunoprecipitation; Ule et al., 2003). CLIP takes advantage of the finding that UV-B irradiation induces covalent complexes between protein-nucleic acids when contact distances are within ~1 Å (Darnell 2010). By applying UV-irradiation to acutely dissected mouse brains, Nova-RNA complexes in situ were stabilized, allowing rigorous purification. After purification through removal of the protein with proteinase K,

CLIP also established that RNA could be efficiently sequenced, using RNA linker ligation and reverse transcription-PCR. Analysis of 34 transcripts identified independently among the first 340 CLIP targets sequenced revealed that they were enriched in YCAY elements, that 71% encoded neuronal synaptic proteins, and that they included Nova-regulated alternative exons (Ule et al. 2003). Hence CLIP was established as a genome-wide unbiased means of identifying functional RNA-protein interaction sites.

To broaden the observations made with CLIP, high throughput sequencing methods, termed HITS-CLIP, were applied to the study of Nova-RNA interactions in 2008 (Licatalosi et al. 2008). In these experiments, millions of Nova-crosslinked RNA fragments of ~50–100nt were sequenced and very conservatively winnowed into a pool of 168,632 unique tags mapping across the mouse genome. These validated predicted Nova-regulated alternative exons that had been correlated with Nova by exon junction array and bioinformatic strategies. They also identified new clusters of Nova tags surrounding alternative exons that were in fact predictive of de novo regulated exons, and they confirmed the position-dependent bioinformatic map. Moreover, these studies also identified many previously unknown Nova binding sites. For example, clusters of Nova-RNA tags within 3' UTRs were found in some cases to surround alternative poly(A) sites, and this finding led to biochemical validation and the discovery that Nova was able to regulate alternative polyadenylation choice in the brain.

These studies were thus able to close the loop on a new approach to genome-wide, unbiased discovery of RBP-RNA function. Application of basic biochemical principles to identify high affinity targets, descriptive analysis of RNA variants under conditions of a defined variable, such as a genetic null, and HITS-CLIP to discriminate direct from indirect target transcripts offer a powerful approach to discovery (Licatalosi and Darnell 2010).

Future directions for this approach are several. The wealth of data coming from HITS-CLIP suggests that new bioinformatic approaches will be important in synthesizing information. Indeed, recent studies have demonstrated the use of Bayesian networks to combine data from such datasets as exon junction arrays, HITS-CLIP, and motif analysis to both validate and produce new discovery regarding RBP-RNA interaction (Zhang et al. 2010). The approaches worked out for Nova provide general solutions. A number of other RBPs have now been studied by HITS-CLIP (Hafner et al. 2010; Konig et al. 2010; Darnell 2010; Sanford et al. 2009; Yeo et al. 2009; Xue et al. 2009) in efforts that, when combined, may provide the groundwork for development of a true combinatorial map of RNA regulation. Finally, HITS-CLIP analysis has been extended to the study of ternary interactions between small non-coding RNAs, mRNA and regulatory proteins. This analysis was first established for the mammalian Ago protein, demonstrating that Ago-mRNA crosslinked tags cluster directly over microRNA binding seeds, such that these ternary HITS-CLIP maps could decode sites of miRNA-mRNA interactions on a robust, genome-wide basis (Chi et al. 2009). Subsequently these studies have been reproduced in C. elegans (Zisoulis et al. 2010) and mammalian tissue culture cells (Hafner et al. 2010) and have begun with other small RNAs (piRNAs; Xu et al. 2009).

These studies offer the exciting possibilities of both developing better understanding and means of capitalizing upon small RNA regulation of gene expression and overlaying such maps with those developed with HITS-CLIP analysis of more traditional RBP-RNA interactions.

Acknowledgments This work is an overview of the efforts of a large number of people and their experiments undertaken over the years. While I have tried to cite the efforts of all, I would especially thank Steven Burley, Ron Buckanovich, Jennifer Darnell, Kate Dredge, Lily Jan, Kirk Jensen, Hal Lewis, Aldo Mele, Kiran Musunuru, Giovanni Stefani, Jernej Ule, Donny Licatalosi, Sung-Wook Chi and Chaolin Zhang for major contributions toward developing the key points in the development of the story told here. This work was supported by the NIH (R01 NS34389 and NS40955 to RBD) and the Howard Hughes Medical Institute. RBD is an Investigator of the Howard Hughes Medical Institute.

References

Albert ML, Darnell RB (2004) Paraneoplastic neurological degenerations: keys to tumour immunity. Nat Rev Cancer 4:36–44

Albert ML, Darnell JC, Bender A, Francisco LM, Bhardwaj N, Darnell RB (1998) Tumor-specific killer cells in paraneoplastic cerebellar degeneration. Nat Med 4:1321–1324

Albert ML, Austin LM, Darnell RB (2000) Detection and treatment of activated T cells in the cerebrospinal fluid of patients with paraneoplastic cerebellar degeneration. [see comments]. Ann Neurol 47:9–17

Buckanovich RJ, Darnell RB (1997) The neuronal RNA binding protein Nova-1 recognizes specific RNA targets in vitro and in vivo. Mol Cell Biol 17:3194–3201

Buckanovich RJ, Posner JB, Darnell RB (1993) Nova, the paraneoplastic Ri antigen, is homologous to an RNA-binding protein and is specifically expressed in the developing motor system. Neuron 11:657–672

Buckanovich RJ, Yang YY, Darnell RB (1996) The onconeural antigen Nova-1 is a neuron-specific RNA-binding protein, the activity of which is inhibited by paraneoplastic antibodies. J Neurosci 16:1114–1122

Chi SW, Zang JB, Mele A, Darnell RB (2009) Argonaute HITS-CLIP decodes microRNA-mRNA interaction maps. Nature 460:479–486

Darnell RB (1996) Onconeural antigens and the paraneoplastic neurologic disorders: at the intersection of cancer, immunity and the brain. Proc Natl Acad Sci USA 93:4529–4536

Darnell RB (2006) Developing global insight into RNA regulation. Cold Spring Harb Symp Quant Biol 71:321–327

Darnell RB (2010) HITS-CLIP: panoramic views of protein-RNA regulation in living cells. Wiley Interdiscip Rev RNA (in press)

Darnell RB, Posner JB (2006) Paraneoplastic syndromes affecting the nervous system. Semin Oncol 33:270–298

Darnell RB, Furneaux H, Posner JB (1989) Characterization of antigens bound by CSF and serum of a patient with cerebellar degeneration: co-expression in Purkinje cells and tumor lines of neuroectodermal origin. Neurology 39:385

Darnell RB, Furneaux HM, Posner JB (1991) Antiserum from a patient with cerebellar degeneration identifies a novel protein in Purkinje cells, cortical neurons, and neuroectodermal tumors. J Neurosci 11:1224–1230

Dredge BK, Darnell RB (2003) Nova regulates GABA(A) receptor gamma2 alternative splicing via a distal downstream UCAU-rich intronic splicing enhancer. Mol Cell Biol 23:4687–4700

Dredge BK, Polydorides AD, Darnell RB (2001) The splice of life: alternative splicing and neurological disease. Nat Rev Neurosci 2:43–50

Dredge BK, Stefani G, Engelhard CC, Darnell RB (2005) Nova autoregulation reveals dual functions in neuronal splicing. EMBO J 24:1608–1620

Graus F, Cordon-Cardo C, Posner JB (1985) Neuronal antinuclear antibody in sensory neuronopathy from lung cancer. Neurology 35:538–543

Green R, Ellington AD, Bartel DP, Szostak JW (1991) In vitro genetic analysis: selection and amplification of rare functional nucleic acids. Meth Compan Methods Enzymol 2:75–86

Hafner M, Landthaler M, Burger L, Khorshid M, Hausser J, Berninger P, Rothballer A, Ascano MJ, Jungkamp AC, Munschauer M, Ulrich A, Wardle GS, Dewell S, Zavolan M, Tuschl T (2010) Transcriptome-wide identification of RNA-binding protein and microRNA target sites by PAR-CLIP. Cell 141:129–141

Hinman MN, Lou H (2008) Diverse molecular functions of Hu proteins. Cell Mol Life Sci 65:3168–3181

Jensen KB, Musunuru K, Lewis HA, Burley SK, Darnell RB (2000) The tetranucleotide UCAY directs the specific recognition of RNA by the Nova K-homology 3 domain. Proc Natl Acad Sci USA 97:5740–5745

Konig J, Zarnack K, Rot G, Curk T, Kayikci M, Blaz Z, Turner DJ, Luscombe NM, Ule J (2010) iCLIP reveals the function of hnRNP particles in splicing at individual nucleotide resolution. Nat Struct Mol Biol 17:909–915

Lewis HA, Musunuru K, Jensen KB, Edo C, Chen H, Darnell RB, Burley SK (2000) Sequence-specific RNA binding by a Nova KH domain: implications for paraneoplastic disease and the fragile X syndrome. Cell 100:323–332

Licatalosi DD, Darnell RB (2006) Splicing regulation in neurologic disease. Neuron 52:93–101

Licatalosi DD, Darnell RB (2010) RNA processing and its regulation: global insights into biological networks. Nat Rev Genet 11:75–87

Licatalosi DD, Mele A, Fak JJ, Ule J, Kayikci M, Chi SW, Clark TA, Schweitzer AC, Blume JE, Wang X, Darnell JC, Darnell RB (2008) HITS-CLIP yields genome-wide insights into brain alternative RNA processing. Nature 456:464–469

McKeever MO, Darnell RB (1992) NAP, a human cerebellar degeneration antigen, is a novel, neuron specific, adaptin family member. Soc Neurosci Abstr 18:1092

Musunuru K, Darnell RB (2001) Paraneoplastic neurologic disease antigens: RNA-binding proteins and signaling proteins in neuronal degeneration. Annu Rev Neurosci 24:239–262

Newman LS, McKeever MO, Okano HJ, Darnell RB (1995) β-NAP, a cerebellar degeneration antigen, is a neuron-specific vesicle coat protein. Cell 82:773–783

Roberts WK, Darnell RB (2004) Neuroimmunology of the paraneoplastic neurological degenerations. Curr Opin Immunol 16:616–622

Sanford JR, Wang X, Mort M, Vanduyn N, Cooper DN, Mooney SD, Edenberg HJ, Liu Y (2009) Splicing factor SFRS1 recognizes a functionally diverse landscape of RNA transcripts. Genome Res 19:381–394

Santomasso BD, Roberts WK, Thomas A, Williams T, Blachere NE, Dudley ME, Houghton AN, Posner JB, Darnell RB (2007) A T-cell receptor associated with naturally occurring human tumor immunity. Proc Natl Acad Sci USA 104:19073–19078

Szabo A, Dalmau J, Manley G, Rosenfeld M, Wong E, Henson J, Posner JB, Furneaux HM (1991) HuD, a paraneoplastic encephalomyelitis antigen contains RNA-binding domains and is homologous to Elav and sex lethal. Cell 67:325–333

Szostak JW, Ellington AD (1993) In vitro selection of functional RNA sequences. Cold Spring Harbor Laboratory Press, Cold Spring Harbor, NY

Ule J, Jensen KB, Ruggiu M, Mele A, Ule A, Darnell RB (2003) CLIP identifies Nova-regulated RNA networks in the brain. Science 302:1212–1215

Ule J, Ule A, Spencer J, Williams A, Hu JS, Cline M, Wang H, Clark T, Fraser C, Ruggiu M, Zeeberg BR, Kane D, Weinstein JN, Blume J, Darnell RB (2005) Nova regulates brain-specific splicing to shape the synapse. Nat Genet 37:844–852

Ule J, Stefani G, Mele A, Ruggiu M, Wang X, Taneri B, Gaasterland T, Blencowe BJ, Darnell RB (2006) An RNA map predicting Nova-dependent splicing regulation. Nature 444:580–586

Xu M, Medvedev S, Yang J, Hecht NB (2009) MIWI-independent small RNAs (MSY-RNAs) bind to the RNA-binding protein, MSY2, in male germ cells. Proc Natl Acad Sci USA 106:12371–12376

Xue Y, Zhou Y, Wu T, Zhu T, Ji X, Kwon YS, Zhang C, Yeo G, Black DL, Sun H, Fu XD, Zhang Y (2009) Genome-wide analysis of PTB-RNA interactions reveals a strategy used by the general splicing repressor to modulate exon inclusion or skipping. Mol Cell 36:996–1006

Yang YY, Yin GL, Darnell RB (1998) The neuronal RNA-binding protein Nova-2 is implicated as the autoantigen targeted in POMA patients with dementia. Proc Natl Acad Sci USA 95:13254–13259

Yeo GW, Coufal NG, Liang TY, Peng GE, Fu XD, Gage FH (2009) An RNA code for the FOX2 splicing regulator revealed by mapping RNA-protein interactions in stem cells. Nat Struct Mol Biol 16:130–137

Zhang C, Frias MA, Mele A, Licatalosi D, Darnell RB (2010) Integrative modeling defines a comprehensive splicing-regulatory network and its combinatorial controls. Science 329:439–443

Zisoulis DG, Lovci MT, Wilbert ML, Hutt KR, Liang TY, Pasquinelli AE, Yeo GW (2010) Comprehensive discovery of endogenous Argonaute binding sites in Caenorhabditis elegans. Nat Struct Mol Biol 17:173–179

Bridging Environment and DNA: Activity-Induced Epigenetic Modification in the Adult Brain

Dengke K. Ma, Junjie U. Guo, Guo-li Ming, and Hongjun Song

Abstract The brain continuously receives sensory information from the outside world and processes the information into electrical activity. Sensory experience in the form of neuronal activity leaves marks in neurons by dynamically modifying neuronal properties, such as connectivity, excitability, gene expression and epigenetic modification. Although DNA methylation has long been considered to be a relatively stable epigenetic marker, recent studies demonstrate that epigenetic modification through changes in DNA methylation can be induced by neuronal activity, learning-related stimuli, and various external cues. Activity-dependent induction of the gene Gadd45b links neuronal activity to DNA demethylation machineries that act in specific loci of the neuronal genome. Thus, a novel mechanism has emerged to bridge the environment and DNA through epigenetic modification of the neuronal DNA methylation landscape, which may have widespread implications for novel mechanisms of neural plasticity and potential therapeutic interventions for neurological and psychiatric disorders.

1 Introduction

Nearly all living organisms have essential capacities to respond and adapt to their environment by actively adjusting themselves, and such plasticity is critical to life. As unicellular organisms with primitive adaptive machineries evolved into more complex forms, the resulting nervous system also acquired more

H. Song (✉)
Institute for Cell Engineering, Departments of Neurology and Neuroscience, Johns Hopkins University School of Medicine, 733 N. Broadway, BRB 759, Baltimore, MD 21205, USA
e-mail: shongju1@jhmi.edu

T. Curran and Y. Christen (eds.), *Two Faces of Evil: Cancer and Neurodegeneration*,
Research and Perspectives in Alzheimer's Disease, DOI 10.1007/978-3-642-16602-0_10,
© Springer-Verlag Berlin Heidelberg 2011

complex sets of molecular, cellular, and circuit-level mechanisms for adaptation. While the architecture of the nervous system is largely specified by genetic programs during neural development, environmental cues and related sensory experience also play essential roles in refining the development of neural circuitry and in forming long-lasting memories and other types of experience-dependent neuronal and behavioral modifications in adult organisms (Katz and Shatz 1996; Zhang and Poo 2001; Pittenger and Kandel 2003; Ming and Song 2005; Flavell and Greenberg 2008; Holtmaat and Svoboda 2009; Ma et al. 2009c). Information from the environment and experience are ultimately processed into electrical activity and integrated by neural networks of the brain to execute adaptive behavior decisions. Research over the past several decades has demonstrated that sensory experience and environmental cues in the form of neuronal activity leave marks in neurons by dynamically modifying neuronal properties, such as excitability, synaptic connectivity, gene expression and epigenetic modifications of mature neurons in specific neuronal circuitry (Pittenger and Kandel 2003; Zhang and Linden 2003; Flavell and Greenberg 2008). Pioneering work in the invertebrate, *Aplysia*, and subsequent work in many other systems have suggested that experience-dependent modification of synaptic strengths serves as a major neuronal substrate for learning and memory (Pittenger and Kandel 2003). The mechanisms by which this happens, however, are exceedingly diverse. Nevertheless, relatively long-lasting effects of adaptive learning, such as formation of long-term memories, invariably require the activation of new gene transcription.

Gene transcription occurs in the context of chromatin, a DNA-protein complex that regulates the accessibility, in space and time, of genes for activation by transcription factors. Such exquisite regulation is implemented by various types of chromatin remodeling factors and chemical group modifications, such as histone acetylation, histone and DNA methylation (Jenuwein and Allis 2001; Jaenisch and Bird 2003; Cedar and Bergman 2009; Yoo and Crabtree 2009). As one major type of chromatin modification, DNA methylation occurs at the 5 position of the DNA cytosine ring. Catalyzed by a conserved family of maintenance methyltransferases (DNMT), 5-cytosine methylation from each individual strand of DNA can be inherited during cell divisions (Bestor and Ingram 1983; Holliday 1993; Hermann et al. 2004). Such DNA sequence-independent, heritable maintenance of gene information is called "epigenetic;" it is considered to be highly stable and represents some sort of "cellular memory" (Holliday 1993; Probst et al. 2009). These intriguing properties led Francis Crick to hypothesize that the unusual stability of long-term memories formed by the brain might be based on the self-perpetuating modification of specific neuronal factors, such as the mechanism for perpetuation of DNA methylation (Crick 1984). Robin Holliday further proposed that specific sites in the neuronal DNA involved in memory might exist in binary methylated or non-methylated states (Holliday 1999). Although these ideas indicated that the prominent presence and modifiable nature of DNA methylation in the brain might play important roles in neural plasticity, concrete experimental evidence was not obtained until recently.

2 DNA Methylation in the Brain

DNA modification with 5-cytosine methylation is widespread among protists, fungi, plants, and animals (Colot and Rossignol 1999; Zemach et al. 2010). Although it is absent in some species and its genomic distribution can vary substantially among different organisms, DNA methylation has been suggested, from an evolutionary perspective, to be important in providing unique possibilities for building functions of various types in different clades of life (Colot and Rossignol 1999). For example, DNA methylation with largely congruent genomic distribution in vertebrates appears to function mainly by silencing genomic activity. Surprisingly, tissue-specific measurements of DNA methylation in both humans and mice revealed markedly high levels of DNA methylation in the brain and suggested an additional level of function for DNA methylation in brain-specific genomic regulation (Ehrlich et al. 1982; Tawa et al. 1990; Inano et al. 2000; Sharma et al. 2005). Analyzing the total base composition of DNA from seven different normal human tissues, Ehrlich et al. (1982) showed that the two most highly methylated DNAs were from thymus and brain, with 1.00 and 0.98 mole percent 5-methylcytosine, respectively. Placental and sperm-derived DNA corresponded to the two least methylated DNAs. Similar results were obtained by Tawa et al. (1990) using mouse tissues. Interestingly, the tissue specificity in the 5-methylcytosine level was observed in the late-fetal stage, and the level continued to change during the subsequent periods, suggesting the importance of the peri- and postnatal periods in the establishment of tissue specificity in DNA methylation content. Notably, such periods match the time window during which experience-driven neuronal changes are the most prominent (Katz and Shatz 1996).

Important roles for DNA methylation in the brain have been further suggested from recent surprising discoveries. The maintenance DNA methyltransferase (DNMT) was once thought to be only important for dividing cells, since epigenetic inheritance of the DNA methylation pattern is no longer needed in post-mitotic cells, for example, mature neurons. Post-mitotic neurons in the postnatal and adult brain, however, possess high levels of Dnmt1, a maintenance methyltransferase (Goto et al. 1994; Inano et al. 2000). Interestingly, Dnmt1 appears to regulate important neural plasticity genes, such as Reelin, and the expression of Dnmt1 itself can be dynamically regulated by a variety of stimuli and psychiatric states (Noh et al. 2005; Guidotti et al. 2007; Dong et al. 2008; Satta et al. 2008; Feng et al. 2010). The de novo DNA methyltransferase Dnmt3a is also highly expressed in non-dividing neurons (Feng et al. 2005). Such observations have been made in both the central nervous system and the olfactory neuroepithelium (MacDonald et al. 2005).

Functionally, DNMTs play important roles in post-mitotic neurons in brain-specific functions. Feng et al. (2010) have recently generated conditional mutant mice lacking either *Dnmt1* or *Dnmt3a* individually or together exclusively in forebrain excitatory neurons. Strikingly, they found that double knockout mice showed abnormal long-term plasticity in the CA1 region of the hippocampus.

Furthermore, the mutant mice exhibited severe behavioral deficits in learning and memory in the standard Morris water maze test. Consistent with a role for DNA methylation in neuronal gene regulation, lack of DNMTs led to substantial DNA demethylation in mutant mice and was correlated with deregulated expression of genes known to be involved in synaptic plasticity. Other studies using normal animals have also provided ample evidence that active modification of DNA, including both DNA methylation and demethylation, can indeed be induced in neurons by various stimuli.

3   Active Modification of DNA Methylation in Neurons

Neurons are excitable cells and are mostly activated by membrane depolarization. Martinowich et al. (2003) showed that membrane depolarization of cultured cortical neurons led to a decrease in CpG methylation within the regulatory region of the brain-derived neurotrophic factor (Bdnf) gene, which is correlated with its increased expression in neurons. The demethylation was observed several hours after depolarization of a post-mitotic neuronal population, suggesting active modes of DNA demethylation. Interestingly, such demethylation appeared to occur in a highly site-specific manner, indicating the functional importance of specific sites with changes in DNA methylation, presumably for regulation of gene transcription. Additional studies showed that treatment of cultured post-mitotic neurons with DNMT inhibitors induced an activity-driven demethylation of promoter I of Bdnf gene, which was mediated by synaptic activation of NMDA (N-methyl-D-aspartic acid) receptors (Nelson et al. 2008).

In a more physiological in vivo system, Weaver et al. (2004) reported that maternal behaviors, such as increased pup licking and grooming and arched-back nursing, by rat mothers altered DNA methylation at a glucocorticoid receptor gene promoter in the hippocampus of their offspring. Similar correlations have been observed in humans (McGowan et al. 2009). Such modification of DNA methylation appeared to be long-lasting and was causally associated with altered histone acetylation and transcription factor binding to the glucocorticoid receptor promoter. To further test whether such epigenetic programming is reversible in adult life, the group infused the adult offspring with the precursor to S-adenosyl-methionine that serves as the donor of methyl groups for DNA methylation and found the reversal of DNA methylation induced by behaviors (Weaver et al. 2005). These results thus suggest that DNA methylation established early in life through behavioral programming is reversible in the adult brain.

Physiological learning-related stimuli also induce active modification of DNA methylation in the adult brain. Miller and Sweatt (2007) showed that contextual fear conditioning, a hippocampus-dependent associative memory paradigm, induced rapid methylation and demethylation of target genes involved in memory formation. Specifically, activity-induced DNA methylation is associated with transcriptional silencing of the memory suppressor gene Protein Phosphatase 1 (PP1), whereas demethylation is associated with transcriptional activation of the synaptic plasticity gene Reelin. The results also implicate highly regulated methyltransferase and demethylase activities during the memory consolidation process. In a related study, Lubin et al. (2008) further showed that contextual fear learning induced differential changes in Bdnf DNA methylation in an exon-specific manner.

To extend these observations and unequivocally establish activity-induced active DNA methylation and demethylation in vivo, we have used electroconvulsive treatment (ECT) in mice to synchronously activate dentate granule neurons, followed by DNA methylation analysis of the homogeneous granule neuron population (Ma et al. 2009a, b). No global DNA demethylation was detected 4 h after ECT in vivo. We next used antibodies against methylated DNA for immunoprecipitation to screen for region-specific DNA demethylation and found significant demethylation at specific regulatory regions of Bdnf and fibroblast growth factor–1 (Fgf-1). Bisulfite sequencing analysis further confirmed ECT-induced demethylation within the regulatory region IX of Bdnf and the brain-specific promoter B of Fgf-1. More physiological stimuli, such as running exercise, and further cell purification for post-mitotic neurons resulted in similar demethylation. The very acute manner (within 4 h) in which such DNA demethylation occurs strongly suggests active DNA modification in post-mitotic mature neurons, at least in the dentate gyrus of the adult brain.

4 Gadd45b Links Neuronal Activity to DNA Demethylation

Numerous observations of active DNA methylation and demethylation in neurons raise an important question about the underlying molecular mechanism. In dividing cells, DNA methylation is maintained through DNMT1 during cell division and formed de novo by DNMT3a/b, whereas demethylation can occur passively through simple loss of methylation markers by restricting maintenance methytransferase activity. Without resorting to cell division, post-mitotic cells appear to use DNA excision repair-like mechanisms to actively remove methylated bases or nucleotides from the DNA directly, or indirectly after modification through deamination or hydroxylation (Metivier et al. 2008; Rai et al. 2008; Ma et al. 2009a, b; Tahiliani et al. 2009). Once the methylated base or nucleotide is removed, the DNA repair machinery can then fill in with an unmethylated base or nucleotide, completing demethylation. Depending on how the methyltransferase activity is regulated in the cell, the newly filled-in base or nucleotide may resume a methylated or demethylated state (Brooks et al. 1996; Ma et al. 2009a, b). Substantial genetic and biochemical studies have provided evidence for such excision repair-like mechanisms of DNA demethylation. Could this be similarly used by neurons? If so, how does neuronal activity impinge on and specifically regulate excision repair processes in neurons?

A clue for these answers came from our recent discovery of the neuronal activity-inducible expression of a gene called Gadd45b (Growth arrest and

DNA-damage-inducible beta) (Ma et al. 2009a,b). Using electroconvulsive treatment to activate mature dentate granule cells in vivo, a paradigm previously used to identify activity-induced immediate early genes and to robustly induce adult hippocampal neurogenesis, we screened for expression changes of known epigenetic regulators by both candidate and microarray experiments. Gadd45b was one of the most highly up-regulated genes identified from the screen. Gadd45b encodes members of a conserved protein family, including Gadd45a, Gadd45b and Gadd45g, that are known to promote both DNA base excision and nucleotide excision repair processes in vitro and in vivo (Smith et al. 1994; Barreto et al. 2007; Jung et al. 2007). Gadd45a has previously been suggested to promote active DNA demethylation in culture by promoting nucleotide excision repair (Barreto et al. 2007). Our experiments showed that mice with Gadd45b deletion exhibited specific deficits in neural activity-induced proliferation of neural progenitors and dendritic growth of newborn neurons in the adult hippocampus. We further showed that neuronal activity-induced Gadd45b promoted activity-induced DNA demethylation of specific promoters and expression of corresponding genes critical for adult neurogenesis, including BDNF and FGFs (Ma et al. 2009a, b). It is interesting to note that Gadd45b corresponds to classic neuronal activity-inducible immediate early gene, such as Arc and cFos. Gadd45b is rapidly but transiently activated in neurons by neuronal activity, including both ECT in vivo and membrane depolarization in vitro. In addition, Gadd45b expression can be induced by light in the suprachiasmatic nucleus and visual cortex (Majdan and Shatz 2006; Porterfield et al. 2007), by learning-related stimuli and by induction of long-term potentiation in vivo (Hevroni et al. 1998; Keeley et al. 2006). Intriguingly, neuronal activity-induced Gadd45b is dependent on NMDA receptor activation and highly co-localized with the classic immediate early gene Arc. This finding suggests that Gadd45b may play cell-autonomous roles in regulating neuronal plasticity, in addition to the character-ized non-cell-autonomous role in regulating activity-induced adult neurogenesis from local neural stem cells (Ma et al. 2009b). Thus, Gadd45b functions as an ideal activity sensor, potentially translating specific pattered neural activities into relatively more stable nuclear changes, such as epigenetic changes in DNA methylation, to elicit long-lasting modifications of neuronal properties and local circuit connectivity (Fig. 1).

How would activity-induced Gadd45b promote its target DNA demethylation? Although direct in vivo evidence for its demethylation mechanism is still lacking, ex vivo and in vitro studies of its related family members have yielded significant insights. Gadd45 proteins are small and acidic and interact with nuclear receptor-type transcription factors and excision repair proteins and specifically bind to acetylated histones at high affinity (Carrier et al. 1999; Yi et al. 2000; Ma et al. 2009a, b), suggesting that, on one hand, the excision repair "demethylase" complex might be loaded to specific genomic loci through Gadd45b-mediated recruiting to transcription factors. On the other hand, the complex could also remove the DNA-histone chromatin barrier through Gadd45b-mediated histone sequestrating to help clear the way for demethylation machineries on methylated DNA during active gene transcription. These two modes are not mutually exclusive. Carrier et al. (1999), for example, showed that Gadd45a can mediate chromatin decondensation through high-affinity interaction with acetylated histones. Gadd45a has also been

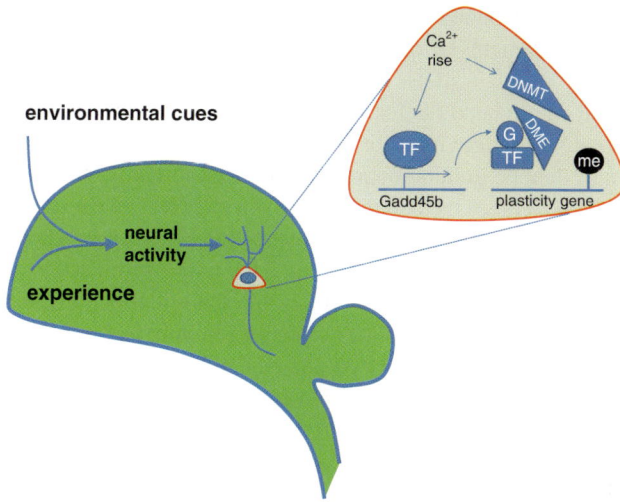

Fig. 1 Gadd45b links neuronal activity to epigenetic modification of neurons. Environmental cues and experience in the form of neural activity modify many aspects of neuronal properties, such as the epigenetic landscape of neurons, as the basis for neural plasticity. DNA methyltransferase (DNMT) functions appear to be under dynamic regulation of neuronal activity. Similar to other immediate early genes, Gadd45b can be induced by neuronal activity in an NMDAR- and calcium-dependent manner. The increased level of Gadd45b protein (G) further promotes epigenetic DNA demethylation of selective genomic targets by facilitating the "demethylase" (DME) function performed by the neuronal DNA excision repair protein complex. The genomic target specificity might be achieved by the interaction of Gadd45b with certain sequence-specific transcription factors (TFs)

shown to facilitate the coupling of DNA deaminases and excision repair enzymes, or enhance the nucleotide excision repair activity in promoting DNA demethylation (Barreto et al. 2007; Rai et al. 2008). Other recent studies also demonstrated that Gadd45 proteins can directly assemble RNA polymerase II transcription machinery sequentially with the nucleotide excision repair (NER) factors at gene promoters (Schmitz et al. 2009; Le May et al. 2010).

5 Neuronal Activity Modifies the DNA Methylation Landscape

Most studies discussed above have shown specific examples of activity-induced changes in DNA methylation. The genomic scope and global properties of activity-induced active DNA modifications in neurons, however, remain largely unknown. Based on an Illumina sequencing-based method for nonbiased, genome-wide analysis at single-nucleotide resolution (Ball et al. 2009), we quantitatively compared the CpG methylation landscape of adult mouse dentate granule neurons in vivo before and after synchronous neuronal activation. We identified a large number of CpGs that exhibit demethylation or de novo methylation within 4 h of neuronal activation. There are two alleles of a given CpG in each cell that can be fully methylated, half-methylated (as for some imprinted genes), or fully unmethylated, and the quantitative changes indicate binary methylation switches at these CpG in a large fraction of dentate granule neurons in vivo. Importantly, we found that some of the methylation changes were maintained even at 24 h after neuronal stimulation, whereas other methylation changes were reversed to basal levels. Clustering analysis shows that the methylation pattern at 24 h after stimulation more closely resembles the pattern at 4 h post-activation than the initial pattern, indicating that activity-induced DNA modifications at selected genomic loci in vivo can be long-lasting. We further show that activity-induced CpG modifications require NMDA receptor activation, occur in post-mitotic neurons, and are correlated with activity-induced gene expression changes. While the genome-scale de novo methylation is blocked by specific DNMT inhibitors, acute genomic site-specific demethylation requires Gadd45b induction. Our study thus demonstrates extensive and specific changes of the DNA methylation landscape of mature neurons in response to neuronal activity and implicates active modification of DNA methylation as a previously underappreciated mechanism for activity-dependent gene regulation in the adult brain.

6 Future Perspectives

Research on neuronal activity-dependent gene regulation in the past has primarily focused on transcription factors and histone-modifying enzymes. Increasingly clear evidence now also points to active modification of DNA with methylation and demethylation as additional important players in regulating activity-dependent gene regulation. Compared with transcription factors, the unique properties of such epigenetic DNA modification indicate that it might orchestrate stimulus-dependent gene expression at different genomic loci and time scales.

Great strides have been made in recent years toward understanding how DNA methylation and activity-induced DNA methylation or demethylation impact neurobiology. Many questions remain to be addressed. We need to know the exact functional role of activity-induced changes in DNA methylation. Is it causally linked to long-lasting changes in gene regulation and required for long-term adaptive responses of the brain, such as memory formation? In addition, gene-body methylation identified from recent epigenomic profiling of DNA methylation appears to be extremely ancient in evolution, yet its functional implication is not at all clear, especially for that occurring in the nervous system. Identification of the in vivo mechanism for activity-induced, Gadd45b-dependent DNA demethylation in neurons will be crucial for further understanding this novel mode of regulation by which transient neuronal stimuli from the environment elicit relatively stable changes in the epigenetic landscape of neurons.

The development of many neurological disorders has been linked to dysregulated epigenetic mechanisms of DNA methylation. Rett syndrome patients, for example, have mutations in MeCP2, which encodes a methyl CpG binding protein. Aberrant DNA methylation of specific DNA sequences, and of those at a genomic level, have also been reported for many neurodegenerative diseases, such as Parkinson's disease, Alzheimer's disease, and Huntington's disease, as well as in many other neurological disorders, such as epilepsy, multiple sclerosis and amyotrophic lateral sclerosis (Robertson 2005; Sananbenesi and Fischer 2009; Urdinguio et al. 2009). Further studies of the activity-dependent pathway leading to changes in neuronal epigenetic DNA modification should afford opportunities for developing novel therapeutics against these disorders.

References

Ball MP, Li JB, Gao Y, Lee JH, LeProust EM, Park IH, Xie B, Daley GQ, Church GM (2009) Targeted and genome-scale strategies reveal gene-body methylation signatures in human cells. Nat Biotechnol 27:361–368

Barreto G, Schafer A, Marhold J, Stach D, Swaminathan SK, Handa V, Doderlein G, Maltry N, Wu W, Lyko F, Niehrs C (2007) Gadd45a promotes epigenetic gene activation by repair-mediated DNA demethylation. Nature 445:671–675

Bestor TH, Ingram VM (1983) Two DNA methyltransferases from murine erythroleukemia cells: purification, sequence specificity, and mode of interaction with DNA. Proc Natl Acad Sci USA 80:5559–5563

Brooks PJ, Marietta C, Goldman D (1996) DNA mismatch repair and DNA methylation in adult brain neurons. J Neurosci 16:939–945

Carrier F, Georgel PT, Pourquier P, Blake M, Kontny HU, Antinore MJ, Gariboldi M, Myers TG, Weinstein JN, Pommier Y, Fornace AJ Jr (1999) Gadd45, a p53-responsive stress protein, modifies DNA accessibility on damaged chromatin. Mol Cell Biol 19:1673–1685

Cedar H, Bergman Y (2009) Linking DNA methylation and histone modification: patterns and paradigms. Nat Rev Genet 10:295–304

Colot V, Rossignol JL (1999) Eukaryotic DNA methylation as an evolutionary device. Bioessays 21:402–411

Crick F (1984) Memory and molecular turnover. Nature 312:101

Dong E, Nelson M, Grayson DR, Costa E, Guidotti A (2008) Clozapine and sulpiride but not haloperidol or olanzapine activate brain DNA demethylation. Proc Natl Acad Sci USA 105:13614–13619

Ehrlich M, Gama-Sosa MA, Huang LH, Midgett RM, Kuo KC, McCune RA, Gehrke C (1982) Amount and distribution of 5-methylcytosine in human DNA from different types of tissues of cells. Nucleic Acids Res 10:2709–2721

Feng J, Chang H, Li E, Fan G (2005) Dynamic expression of de novo DNA methyltransferases Dnmt3a and Dnmt3b in the central nervous system. J Neurosci Res 79:734–746

Feng J, Zhou Y, Campbell SL, Le T, Li E, Sweatt JD, Silva AJ, Fan G (2010) Dnmt1 and Dnmt3a maintain DNA methylation and regulate synaptic function in adult forebrain neurons. Nat Neurosci 13:423–430

Flavell SW, Greenberg ME (2008) Signaling mechanisms linking neuronal activity to gene expression and plasticity of the nervous system. Annu Rev Neurosci 31:563–590

Goto K, Numata M, Komura JI, Ono T, Bestor TH, Kondo H (1994) Expression of DNA methyltransferase gene in mature and immature neurons as well as proliferating cells in mice. Differentiation 56:39–44

Guidotti A, Ruzicka W, Grayson DR, Veldic M, Pinna G, Davis JM, Costa E (2007) S-adenosyl methionine and DNA methyltransferase-1 mRNA overexpression in psychosis. NeuroReport 18:57–60

Hermann A, Goyal R, Jeltsch A (2004) The Dnmt1 DNA-(cytosine-C5)-methyltransferase methylates DNA processively with high preference for hemimethylated target sites. J Biol Chem 279:48350–48359

Hevroni D, Rattner A, Bundman M, Lederfein D, Gabarah A, Mangelus M, Silverman MA, Kedar H, Naor C, Kornuc M, Hanoch T, Seger R, Theill LE, Nedivi E, Richter-Levin G, Citri Y (1998) Hippocampal plasticity involves extensive gene induction and multiple cellular mechanisms. J Mol Neurosci 10:75–98

Holliday R (1993) Epigenetic inheritance based on DNA methylation. EXS 64:452–468

Holliday R (1999) Is there an epigenetic component in long-term memory? J Theor Biol 200:339–341

Holtmaat A, Svoboda K (2009) Experience-dependent structural synaptic plasticity in the mammalian brain. Nat Rev Neurosci 10:647–658

Inano K, Suetake I, Ueda T, Miyake Y, Nakamura M, Okada M, Tajima S (2000) Maintenance-type DNA methyltransferase is highly expressed in post-mitotic neurons and localized in the cytoplasmic compartment. J Biochem 128:315–321

Jaenisch R, Bird A (2003) Epigenetic regulation of gene expression: how the genome integrates intrinsic and environmental signals. Nat Genet 33(Suppl):245–254

Jenuwein T, Allis CD (2001) Translating the histone code. Science 293:1074–1080

Jung HJ, Kim EH, Mun JY, Park S, Smith ML, Han SS, Seo YR (2007) Base excision DNA repair defect in Gadd45a-deficient cells. Oncogene 26:7517–7525

Katz LC, Shatz CJ (1996) Synaptic activity and the construction of cortical circuits. Science 274:1133–1138

Keeley MB, Wood MA, Isiegas C, Stein J, Hellman K, Hannenhalli S, Abel T (2006) Differential transcriptional response to nonassociative and associative components of classical fear conditioning in the amygdala and hippocampus. Learn Mem 13:135–142

Le May N, Mota-Fernandes D, Velez-Cruz R, Iltis I, Biard D, Egly JM (2010) NER factors are recruited to active promoters and facilitate chromatin modification for transcription in the absence of exogenous genotoxic attack. Mol Cell 38:54–66

Lubin FD, Roth TL, Sweatt JD (2008) Epigenetic regulation of BDNF gene transcription in the consolidation of fear memory. J Neurosci 28:10576–10586

Ma DK, Guo JU, Ming GL, Song H (2009a) DNA excision repair proteins and Gadd45 as molecular players for active DNA demethylation. Cell Cycle 8:1526–1531

Ma DK, Jang MH, Guo JU, Kitabatake Y, Chang ML, Pow-Anpongkul N, Flavell RA, Lu B, Ming GL, Song H (2009b) Neuronal activity-induced Gadd45b promotes epigenetic DNA demethylation and adult neurogenesis. Science 323:1074–1077

Ma DK, Kim WR, Ming GL, Song H (2009c) Activity-dependent extrinsic regulation of adult olfactory bulb and hippocampal neurogenesis. Ann NY Acad Sci 1170:664–673

MacDonald JL, Gin CS, Roskams AJ (2005) Stage-specific induction of DNA methyltransferases in olfactory receptor neuron development. Dev Biol 288:461–473

Majdan M, Shatz CJ (2006) Effects of visual experience on activity-dependent gene regulation in cortex. Nat Neurosci 9:650–659

Martinowich K, Hattori D, Wu H, Fouse S, He F, Hu Y, Fan G, Sun YE (2003) DNA methylation-related chromatin remodeling in activity-dependent BDNF gene regulation. Science 302:890–893

McGowan PO, Sasaki A, D'Alessio AC, Dymov S, Labonte B, Szyf M, Turecki G, Meaney MJ (2009) Epigenetic regulation of the glucocorticoid receptor in human brain associates with childhood abuse. Nat Neurosci 12:342–348

Metivier R, Gallais R, Tiffoche C, Le Peron C, Jurkowska RZ, Carmouche RP, Ibberson D, Barath P, Demay F, Reid G, Benes V, Jeltsch A, Gannon F, Salbert G (2008) Cyclical DNA methylation of a transcriptionally active promoter. Nature 452:45–50

Miller CA, Sweatt JD (2007) Covalent modification of DNA regulates memory formation. Neuron 53:857–869

Ming GL, Song H (2005) Adult neurogenesis in the mammalian central nervous system. Annu Rev Neurosci 28:223–250

Nelson ED, Kavalali ET, Monteggia LM (2008) Activity-dependent suppression of miniature neurotransmission through the regulation of DNA methylation. J Neurosci 28:395–406

Noh JS, Sharma RP, Veldic M, Salvacion AA, Jia X, Chen Y, Costa E, Guidotti A, Grayson DR (2005) DNA methyltransferase 1 regulates reelin mRNA expression in mouse primary cortical cultures. Proc Natl Acad Sci USA 102:1749–1754

Pittenger C, Kandel ER (2003) In search of general mechanisms for long-lasting plasticity: Aplysia and the hippocampus. Philos Trans R Soc Lond B Biol Sci 358:757–763

Porterfield VM, Piontkivska H, Mintz EM (2007) Identification of novel light-induced genes in the suprachiasmatic nucleus. BMC Neurosci 8:98

Probst AV, Dunleavy E, Almouzni G (2009) Epigenetic inheritance during the cell cycle. Nat Rev Mol Cell Biol 10:192–206

Rai K, Huggins IJ, James SR, Karpf AR, Jones DA, Cairns BR (2008) DNA demethylation in zebrafish involves the coupling of a deaminase, a glycosylase, and gadd45. Cell 135:1201–1212

Robertson KD (2005) DNA methylation and human disease. Nat Rev Genet 6:597–610

Sananbenesi F, Fischer A (2009) The epigenetic bottleneck of neurodegenerative and psychiatric diseases. Biol Chem 390:1145–1153

Satta R, Maloku E, Zhubi A, Pibiri F, Hajos M, Costa E, Guidotti A (2008) Nicotine decreases DNA methyltransferase 1 expression and glutamic acid decarboxylase 67 promoter methylation in GABAergic interneurons. Proc Natl Acad Sci USA 105:16356–16361

Schmitz KM, Schmitt N, Hoffmann-Rohrer U, Schafer A, Grummt I, Mayer C (2009) TAF12 recruits Gadd45a and the nucleotide excision repair complex to the promoter of rRNA genes leading to active DNA demethylation. Mol Cell 33:344–353

Sharma RP, Grayson DR, Guidotti A, Costa E (2005) Chromatin, DNA methylation and neuron gene regulation–the purpose of the package. J Psychiatry Neurosci 30:257–263

Smith ML, Chen IT, Zhan Q, Bae I, Chen CY, Gilmer TM, Kastan MB, O'Connor PM, Fornace AJ Jr (1994) Interaction of the p53-regulated protein Gadd45 with proliferating cell nuclear antigen. Science 266:1376–1380

Tahiliani M, Koh KP, Shen Y, Pastor WA, Bandukwala H, Brudno Y, Agarwal S, Iyer LM, Liu DR, Aravind L, Rao A (2009) Conversion of 5-methylcytosine to 5-hydroxymethylcytosine in mammalian DNA by MLL partner TET1. Science 324:930–935

Tawa R, Ono T, Kurishita A, Okada S, Hirose S (1990) Changes of DNA methylation level during pre- and postnatal periods in mice. Differentiation 45:44–48

Urdinguio RG, Sanchez-Mut JV, Esteller M (2009) Epigenetic mechanisms in neurological diseases: genes, syndromes, and therapies. Lancet Neurol 8:1056–1072

Weaver IC, Cervoni N, Champagne FA, D'Alessio AC, Sharma S, Seckl JR, Dymov S, Szyf M, Meaney MJ (2004) Epigenetic programming by maternal behavior. Nat Neurosci 7:847–854

Weaver IC, Champagne FA, Brown SE, Dymov S, Sharma S, Meaney MJ, Szyf M (2005) Reversal of maternal programming of stress responses in adult offspring through methyl supplementation: altering epigenetic marking later in life. J Neurosci 25:11045–11054

Yi YW, Kim D, Jung N, Hong SS, Lee HS, Bae I (2000) Gadd45 family proteins are coactivators of nuclear hormone receptors. Biochem Biophys Res Commun 272:193–198

Yoo AS, Crabtree GR (2009) ATP-dependent chromatin remodeling in neural development. Curr Opin Neurobiol 19:120–126

Zemach A, McDaniel IE, Silva P, Zilberman D (2010) Genome-Wide Evolutionary Analysis of Eukaryotic DNA Methylation. Science 328:916–919

Zhang W, Linden DJ (2003) The other side of the engram: experience-driven changes in neuronal intrinsic excitability. Nat Rev Neurosci 4:885–900

Zhang LI, Poo MM (2001) Electrical activity and development of neural circuits. Nat Neurosci 4 (Suppl):1207–1214

Intrinsic Brain Signaling Pathways: Targets of Neuron Degeneration

Harry T. Orr

Abstract Cancer and neurodegenerative disease are two of the greatest worries facing the aging population. Alteration in cell signaling pathways is a molecular process that underlies many cancers and is the targets of therapies. Recent data indicate that these pathways also have a key role in neurodegenerative disease. Glutamine tract expansion triggers nine neurodegenerative diseases most likely by conferring toxic properties to the mutant protein. Spinocerebellar ataxia type 1 (SCA1) is typically a late-onset fatal autsosomal dominant neurodegenerative disease characterized by loss of motor coordination and balance. A characteristic feature of SCA1 pathology is atrophy and loss of Purkinje cells (PCs) from the cerebellar cortex. In SCA1, phosphorylation of ataxin-1 (ATXN1) at Ser776 is one biochemical feature thought to be key for pathogenesis. We found that replacing Ser776 with a phospho-mimicking Asp amino acid converts ATXN1 with a wild type glutamine tract, ATXN1[30Q]-D776, into a pathogenic protein. Yet, in contrast to disease induced by ATXN1[82Q] that progresses from dysfunction to cell death, ATXN1[30Q]-D776-induced disease failed to progress beyond dysfunction. These results support a disease model where disease initiation and subsequent neuronal dysfunction are distinct from progression to cell death. Ser776, presumably its phosphorylation, is critical for the pathogenic pathway leading to neuronal dysfunction, while an expanded polyglutamine tract is essential for further progression to neuronal death. A cerebellar extract-based phosphorylation assay revealed that S776 phosphorylation is regulated by the kinase PKA and the phosphatase PP2A. These date form the basis of a small molecule inhibitor screen that could provide an effective treatment for SCA1.

H.T. Orr
Department of Laboratory Medicine and Pathology, Institute of Translational Neuroscience,
University of Minnesota, Minneapolis, MN 55455, USA
e-mail: orrxx002@umn.edu

T. Curran and Y. Christen (eds.), *Two Faces of Evil: Cancer and Neurodegeneration*,
Research and Perspectives in Alzheimer's Disease, DOI 10.1007/978-3-642-16602-0_11,
© Springer-Verlag Berlin Heidelberg 2011

1 Introduction

As described by the other contributors to this volume, two of the greatest fears facing the aging population are the cancers and neurodegenerative diseases. Perhaps because cancer is caused by uncontrolled cell growth and neurodegenerative disease is the result of cellular dysfunction and subsequent atrophy, it is an unexpected finding that the molecular pathways underlying these two major disorders overlap to an extensive degree. No group of neurodegenerative disorders illustrates a link with cancer better than a group of ataxias caused by mutations in genes encoding components of the DNA repair pathway. In dividing cells, the link between DNA repair and unregulated cell division, i.e. cancer, dates back to studies on xeroderma pigmatosum published in 1968 by James E. Cleaver (Cleaver 1968). Postmitotic neurons, in contrast, respond to deficits in DNA repair, with the induction of neuronal cell death that often presents clinically with ataxia (Rass et al. 2007). Among the neurodegenerative ataxias associated with DNA repair deficits are ataxia telangiectasia, ataxia with ocular aprexia Type 1 and spinocerebellar ataxia and neuropathy (Paulson and Miller 2005). What about neurodegenerative disorders that do not present with cancer? Do intrinsic, cancer-related signaling pathways contribute to their pathogenesis?

2 Phosphorylation in SCA1: An Intrinsic Pathway that Drives Pathogenesis

Spinocerebellar ataxia type 1 (SCA1) is one of nine inherited neurodegenerative disorders caused by the expansion of a CAG trinucleotide repeat encoding a polyglutamine tract (Orr and Zoghbi 2007). Besides SCA1, the group of polyglutamine disorders includes spinobulbar muscular atrophy, Huntington's disease (HD), dentatorubral-pallidoluysian atrophy, and SCAs 2, 3, 6, 7, and 17. Genetic evidence indicates that the disease mutation induces a toxic gain-of-function in the mutant polyglutamine protein. The discovery that residues outside of the polyglutamine tract in these proteins are crucial for pathogenesis hints that alterations in the normal function of these proteins are linked to toxicity (Klement et al. 1998; Katsuno et al. 2002; Emamian et al. 2003; Tsuda et al. 2005; Graham et al. 2006; Gu et al. 2009).

Protein phosphorylation is a widespread post-translational modification used to regulate protein function and signal transduction. It affects every basic cellular process, including those associated with tumorgenesis and cell cycle control. A hint that phosphorylation may have a role in a polyglutamine disease came with the finding that serine (Ser/S) 776 in ataxin-1, the protein mutated in SCA1, is normally phosphorylated. Intriguingly, replacing S776 with a non-phosphorylatable alanine (Ala/A) dramatically reduced the toxicity of Ataxin-1 with an expanded polyglutamine tract (Emamian et al. 2003). By both pathological (Fig. 1) and neurological

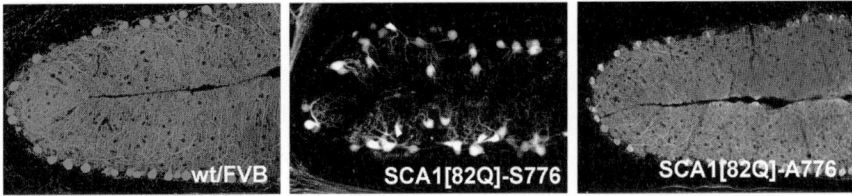

Fig. 1 A phospho-resistant alanine amino acid at position 776 dramatically reduces toxicity in vivo of Ataxain-1 with an expanded glutamine tract. Shown are calbindin, a Purkinje cell-specific marker, immunostained sections of the cerebellar cortex from SCA1 mice at 37 weeks of age

criteria, disease in Ataxin-1[82Q]-A776 mice is essentially undetectable compared to disease in age-matched Atxain-1[82Q]-S776 animals. Moreover, in a *Drosophila* model of SCA1, disease severity directly correlates with levels of the kinase Akt predicted to phosphorylate S776 (Chen et al. 2003).

Subsequently, it was found that the majority of soluble wild type Ataxin-1 and Ataxin-1 with an expanded polyglutamine tract is assembled into large, stable complexes containing regulators of transcription and RNA splicing (Lam et al. 2006; Lim et al. 2008). It seems that both the phosphorylation state of S776 and length of the polyglutamine tract determine which nuclear components assemble with Atxain-1 into high molecular weight complexes. Expansion of the glutamine tract decreases the interaction of Ataxin-1 with the transcriptional regulator Capicua and increases its interaction with the regulator of splicing RBM17. In addition to being affected by the length of the glutamine tract, the Ataxin-1/RBM17 interaction is modulated by the amino acid at residue 776 in Ataxin-1 in a fashion that seems to reflect its phosphorylation state. An Ala at position 776 dramatically reduces the assembly of Ataxin-1/RBM17 complexes, whereas a phosphomimetic aspartic acid (Asp/D) at 776 enhances complex formation (Lim et al. 2008). The Asp at position 776 effect is such that, when placed in a wild type allele of Ataxin-1 with 30 glutamines, its interaction with RBM17 is enhanced to a level comparable to that seen with Ataxin-1[82Q]. Thus, by a single amino acid change, placing a phosphomimetic Asp at residue 776 converts a wild type form of Ataxin-1 into a protein having a biochemical property characteristic of a mutant disease-causing form of the protein. This finding raises the intriguing question of whether in vivo Ataxin-1[30Q]-D776 shares pathogenic properties of Ataxin-1[82Q]. Besides inhibiting the interaction of Ataxin-1 with RBM17, a phospho-resistant Ala at residue 776 destabilizes the protein in tissue culture cells and in vivo in Purkinje cells (Chen et al. 2003; Jorgensen et al. 2009). This effect is due to a disruption of an interaction of Ataxin-1 with the chaperone 14-3-3 (Chen et al. 2003). Thus, phosphorylation of Ataxin-1 at S776 also promotes protein interactions that function to stabilize the protein.

The data support a model wherein Ataxin-1 assembles into at least two native complexes, one with the transcriptional regulator Capicua and a second complex

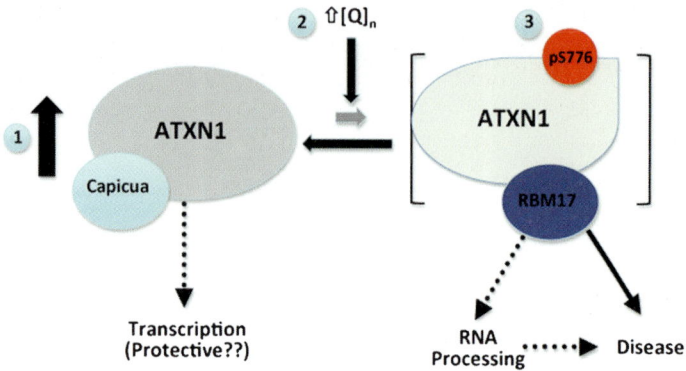

Fig. 2 A model depicting a mechanism by which phosphorylation of S776 in mutant Ataxin-1 promotes disease. Ataxin-1 intrinsically exists in equilibrium between two conformations. One confirmation assembles in a complex with Capicua and another with RBM17. Disease severity is determined by the amount of the RBM17 complex in a neuron. There are three potential means by which the amount of the RBM17 complex can be increased: (1) increase the expression of Ataxin-1, (2) increase the number of glutamines that shift the equilibrium in favor of RBM17 complexes, and (3) increase phosphorylation of S776, which decreases the rate of converting RBM17 complexes to Capicua complexes

with the RNA-binding protein RBM17 (Fig. 2). With glutamine tract expansion and phosphorylation at S776 formation of the complex with RBM17 complex is enhanced. In the Drosophila model of SCA1, reduction of RBM17 is protective against neurodegeneration, whereas an increase in RBM17 promotes degeneration (Lim et al. 2008), further supporting the concept that it is the Ataxnin-1/RBM17 complex that drives pathogenesis. To date, two functions have been ascribed to RBM17 and its Drosophila homologue SPF45: RNA splicing and DNA repair (Chaouki and Salz 2006). Whether one or the other of these pathways is involved in the etiology of SCA1 awaits the results of further studies. Regardless, a treatment targeted at S776 phosphorylation is likely to have a major impact clinically in SCA1. Thus, understanding the pathways that regulate S776 phosphorylation becomes critical.

Sequence analysis of residues surrounding S776 in Ataxin-1 reveals Akt and PKA as the two kinases with highest probability of phosphorylating S776. Inhibiting Akt activity either in vivo or in a cerebellar extract fails to decrease the amount of phospho-S776-ATXN1 (Jorgensen et al. 2009). However, several lines of evidence indicate that PKA is the S776 kinase, including co-fractionation with the peak of S776 phosphorylating activity in cerebellar cytosol and reduction of S776 phosphorylation by cerebellar lysates upon addition of two PKA inhibitors and following PKA immunodepletion.

A cerebellar signaling pathway linked to regulating PKA activity is the rebound potentiation, a form of synaptic plasticity at GABAergic synapses between inhibitory interneurons (basket and stellate neurons) and Purkinje cells. A model described by Kawaguchi and Hirano (2002) assumed that there is a relatively high basal level of PKA activity in Purkinje cells that is reduced with the activation of GABA$_B$R during depolarization at inhibitory synapses. Since phosphorylation of ATXN1 at S776 promotes neurodegeneration (Emamian et al. 2003; Chen et al. 2003), it is tempting to speculate that a high basal activity of PKA in Purkinje cells may contribute to their enhanced sensitivity to the toxic effects of mutant ATXN1 in SCA1.

3 Phosphorylation in Some Other Polyglutamine Neurodegenerative Diseases

In contrast to S776 phosphorylation in Ataxin-1, which seems to increase toxicity, phosphorylation of two other polyglutamine proteins, huntingtin (Huntington's disease) and the androgen receptor (AR; spinal and bulbar muscular atrophy), appears to be protective and decrease toxicity (Pennuto et al. 2009). Huntingtin has been shown to be phosphorylated at multiple serine residues (Humbert et al. 2002; Schilling et al. 2006; Gu et al. 2009). In all cases examined, phosphorylation was associated with reduced toxicity of huntingtin with an expanded polyglutamine tract. In perhaps the most extensive in vivo, study Gu et al. (2009) showed that substitution of phosphomimetic Asp for a Ser at residues 13 and 16 of mutant huntingtin abolished the ability of the protein to induce disease, whereas substitution of phospho-resistant alanines at these positions preserved toxicity in transgenic mice. Like huntingtin, several sites of phosphorylation are known (Gioeli et al. 2002). The biological effect of these phosphorylations, at least in cell lines, appears to vary depending on serine. For example, phosphorylation of S516 increases toxicity (LaFevre-Bernt and Ellerby 2003), whereas phosphorylation at serines 215 and 792 decreases toxicity (Palazzolo et al. 2007). To complicate matters further, phosphorylation at serines 426 and 516 appears to have opposite effects on wild type and mutant AR. Loss of a phosphorylation site at positions 426 and 516 converts wild type AR into a toxic protein, whereas loss of this same site in mutant AR with an expanded glutamine tract reduces toxicity (Funderburk et al. 2009).

4 Perspective

The polyglutamine disorders nicely illustrate that cellular signaling pathways can impact disease course. While the specific effects on disease seem to vary between diseases and, in some cases, between sites of phosphorylation within a protein, dissecting the role of these modifications is likely to lead to novel targets for therapeutic development. Critical in this effort will be identifying the cellular pathway and specific kinase that regulate phosphorylation in vivo. Although this identification has yet to be accomplished for any of the polyglutamine proteins, the putative candidates (Akt, PKA, CDK5, and MAPK) are all well known to those studying cellular signaling pathways in cancer, offering hope that the wealth of drug development for these targets as potential therapies in cancer will accelerate their study as targets in the polyglutamine neurodegenerative disease.

References

Chaouki AS, Salz HK (2006) *Drosophila* SPF45: a bifunctional protein with roles in both splicing and DNA repair. PLoS Genet 2:1974–1983

Chen H-K, Fernandez-Funez P, Acevedo SF, Lam YC, Kaytor MD, Fernandez MH, Aitken A, Skoulakis EMC, Orr HT, Botas J, Zoghbi HY (2003) Interaction of Akt-phosphorylated ataxin-1 with 14-3-3 mediates neurodegeneration in spinocerebellar ataxia type 1. Cell 113:457–468

Cleaver JE (1968) Defective DNA repair replication in xeroderma pigmentosum. Nature 218: 652–656

Emamian ES, Kaytor MD, Duvick LA, Zu T, Tousey SK, Zoghbi HY, Clark HB, Orr HT (2003) Serine 776 of ataxin-1 is critical for polyglutamine-induced disease in SCA1 transgenic mice. Neuron 38:375–387

Funderburk SF, Shatkina L, Mink S, Weis Q, Weg-Remers S, Cato AC (2009) Specific N-terminal mutations in the human androgen receptor induce cytotoxicity. Neurobiol Aging 30: 1851–1864

Gioeli D, Ficarro SB, Kwiek JJ, Aaronson D, Hancock M, Catling AD, White FM, Christian RE, Settlage RE, Shabanowitz J, Hunt DF, Weber MJ (2002) Androgen receptor phosphorylation. Regulation and identification of the phosphorylation sites. J Biol Chem 277:29304–29314

Graham RK, Deng Y, Slow EJ, Haigh B, Bissada N, Lu G, Pearson J, Shehadeh J, Bertram L, Murphy Z, Warby SC, Doty CN, Roy S, Wellington CL, Leavitt BR, Raymond LA, Nicholson DW, Hayden MR (2006) Cleavage at the caspase-6 site is required for neuronal dysfunction and degeneration due to mutant huntingtin. Cell 125:1179–1191

Gu X, Greiner ER, Mishra R, Kodali R, Osmand A, Finkbeiner S, Steffan JS, Thompson LM, Wetzel R, Yang XW (2009) Serines 13 and 16 are critical determinants of full-length mutant huntingtin induced disease pathogenesis in HD mice. Neuron 64:828–840

Humbert S, Bryson EA, Cordelieres FP, Connors NC, Datta SR, Finkbeiner S, Greenberg ME, Saudou F (2002) The IGF-1/Akt pathway is neuroprotective in Huntington's disease and involves Huntingtin phosphorylation by Akt. Dev Cell 2:831–837

Jorgensen ND, Andresen JM, Lagalwar S, Armstrong B, Stevens S, Byam CE, Duvick LA, Lai S, Zoghbi HY, Clark HB, Orr HT (2009) Phosphorylation of ATXN1 at Ser776 in the cerebellum. J Neurochem 110:675–686

Katsuno M, Adachi H, Kume A, Li M, Nakagomi Y, Niwa H, Sang C, Kobayashi Y, Doyu M, Sobue G (2002) Testosterone reduction prevents phenotypic expression in a transgenic mouse model of spinal and bulbar muscular atrophy. Neuron 35:843–854

Kawaguchi S, Hirano T (2002) Signaling cascade regulating long-term potentiation of GABAA receptor responsiveness in cerebellar Purkinje neurons. J Neurosci 22:3969–3976

Klement IA, Skinner PJ, Kaytor MD, Yi H, Hersch SM, Clark HB, Zoghbi HY, Orr HT (1998) Ataxin-1 nuclear localization and aggregation: role in polyglutamine-induced disease in SCA1 transgenic mice. Cell 95:41–53

LaFevre-Bernt MA, Ellerby LM (2003) Kennedy's disease. Phosphorylation of the polyglutamine-expanded form of androgen receptor regulates its cleavage by caspase-3 and enhances cell death. J Biol Chem 278:34918–34924

Lam YC, Bowman AB, Jafar-Neja P, Lim J, Richman R, Fryer JD, Hyun E, Duvick LA, Orr HT, Botas J, Zoghbi HY (2006) Mutant ATAXIN-1 interacts with the repressor Capicua in its native complex to cause SCA1 neuropathology. Cell 127:1335–1347

Lim J, Crespo-Barreto J, Jafar-Nejad P, Bowman AB, Richman R, Hill DE, Orr HT, Zoghbi HY (2008) Opposing effects of polyglutamine expansion on native protein complexes contribute to SCA1. Nature 452:713–719

Orr HT, Zoghbi HY (2007) Trinucleotide repeat disorders. Ann Rev Neurosci 30:575–621

Palazzolo I, Burnett BG, Young JE, Brenne PL, La Spada AR, Fischbeck KH, Howell BW, Pennuto M (2007) Akt blocks ligand binding and protects against expanded polyglutamine androgen receptor toxicity. Human Mol Genet 16:1593–1603

Paulson HL, Miller VM (2005) Breaks in coordination: DNA repair in inherited ataxia. Neuron 46:845–848

Pennuto M, Palazzolo I, Poletti A (2009) Post-translational modifications of expanded polyglutamine proteins: impact on neurotoxicity. Human Mol Genet 18:R40–R47

Rass U, Ahel I, West SC (2007) Defective DNA repair and neurodegenerative disease. Cell 130:991–1004

Schilling B, Gafni J, Torcassi C, Cong X, Row RH, LaFevre-Bernt MA, Cusack MP, Ratovitski T, Hirschhorn R, Ross CA, Gibson BW, Ellerby LM (2006) Huntingtin phosphorylation sites mapped by mass spectrometry. Modulation of cleavage and toxicity. J Biol Chem 281:23686–23697

Tsuda H, Jafar-Nejad H, Patel AJ, Sun Y, Chen HK, Rose MF, Venken KJ, Botas J, Orr HT, Bellen HJ, Zoghbi HY (2005) The AXH domain of Ataxin-1 mediates neurodegeneration through its interaction with Gfi-1/Senseless proteins. Cell 122:633–644

The miRNA System: Bifurcation Points of Cancer and Neurodegeneration

Kenneth S. Kosik, Pierre Neveu, and Sourav Banerjee

Abstract Developmental programs direct cells toward specific fates by harnessing epigenetic mechanisms to progressively limit their potential and canalize changes in cell state toward a specialized terminal identity. Controlling proliferation on the one hand, and apoptosis on the other, are among the paramount twin hazards faced by the cell over its life span. Loosening of controls over proliferation clearly leads to cancer and loosening of controls over apoptosis may lead to degeneration. The genetic logic circuits that regulate proliferation and apoptosis lie at the core of cellular function, and it follows from the complexity of this circuitry that many different pathways exist in nearly all cells that can lead to cancer or degeneration. In neurons, microRNAs (miRNAs) have assumed a specialized role in regulating synaptic plasticity locally at synapses. This emerging role for miRNAs introduces another candidate pathway in the exploration of the underlying molecular mechanisms of neurodegeneration.

1 Introduction

microRNA (miRNA) biology has become recognized as an important basic mechanism in cellular function and dysfunction. The widespread presence of the miRNA system across metazoa points to a fundamental role in the metabolism of cells. The profile of the expressed miRNA population can reveal a highly accurate picture of cell identity. Their profiles unambiguously distinguish between cellular phenotypes, such as embryonic stem cells, a vast variety of precursor cells, terminally differentiated cells including neurons, and tumor types, even among closely related cancers. The post-transcriptional cytoplasmic role of these small RNAs in establishing and maintaining the proteome complements the nuclear role

K.S. Kosik (✉)
Department of Molecular Cellular Developmental Biology, Neuroscience Research Institute, University of California, Santa Barbara, CA 93106, USA
e-mail: kosik@lifesci.ucsb.edu

T. Curran and Y. Christen (eds.), *Two Faces of Evil: Cancer and Neurodegeneration*, Research and Perspectives in Alzheimer's Disease, DOI 10.1007/978-3-642-16602-0_12, © Springer-Verlag Berlin Heidelberg 2011

of transcription factors in determining the transcriptional profile. miRNAs often function as control points in a variety of transcriptional feedback loops, operate according to a mechanistic complexity that rivals transcriptional control, and reveal logic circuits governed more by networks than by a master switch.

In mammals, miRNAs have been reported to have a role in embryogenesis and stem cell maintenance (Bernstein et al. 2003), hematopoietic cell differentiation (Chen et al. 2004), and brain development (Miska 2005; Miska et al. 2004). Because both cancer and neurodegeneration disrupt cell homeostasis and pervasively deregulate cellular systems, exploring miRNAs in these conditions is clearly worthwhile. Although an extensive literature describes dysregulated miRNAs in cancer and a far fewer number of papers describe dysregulated miRNAs in various degenerative conditions (Hébert and De Strooper 2009), the significance of these observations to the underlying pathogenesis remains obscure. Importantly, for all disease research, the difficulty is distinguishing processes that initiate disease from those that are involved in disease progression.

The first miRNA was discovered by Victor Ambros and colleagues Rosalind Lee and Rhonda Feinbaum in 1993 in a screen for genes involved in developmental timing (Lee et al. 1993). This screen revealed a 22-nucleotide non-coding RNA called *lin-4*. Although seven years elapsed until the second miRNA, *let-7* was discovered, a windfall of miRNAs was reported from many species shortly after the discovery of *let-7*, at which time they were termed miRNAs (Lagos-Quintana et al. 2001; Lau et al. 2001; Lee and Ambros 2001). MiRNAs are now considered a key layer of post-transcriptional control within the networks of gene regulation. Nevertheless, despite extensive data that correlate miRNA expression with important cellular processes, such as proliferation and apoptosis, as well as the extraordinary conservation of many miRNAs, individual miRNAs are dispensable.

The biogenesis of miRNAs has been reviewed in many places (Kosik 2006). Cleavage of the primary transcript occurs in the nucleus with a ribonuclease enzyme called Drosha, and cleavage of the precursor to the mature miRNA occurs in the cytoplasm with another ribonuclease called Dicer (Lee et al. 2003; Bernstein et al. 2001; Hutvagner et al. 2001). MiRNAs recognize their targets in the RNA-induced silencing complex (RISC) based on sequence complementarity (Brennecke et al. 2005); however, the complementarity is imperfect. Imperfect pairing is essential for miRNAs because it allows the miRNA to target multiple mRNAs with differential affinities. Once paired, the miRNA will either induce the degradation of the mRNA, impair its translation, or both. The mechanisms of this inhibition have not been fully worked out (Jackson and Standart 2007; Pillai et al. 2007). Although it has been written that, in humans, miRNAs usually inhibit protein translation of their target genes and only infrequently cause degradation or cleavage of the messenger RNA (Bartel 2004), in fact, small changes in mRNA levels of target genes are frequently observed. However, some of these changes may not be due to the degradation of the target mRNA by the miRNA; instead, target mRNAs may undergo upstream inhibition coincident with direct suppression by the miRNA.

2   miRNAs and Cancer

The key early observations that implicated miRNAs in cancer included: (1) among the first miRNAs discovered in *C. elegans* and *Drosophila* were some that controlled cell proliferation and apoptosis (Brennecke et al. 2003; Lee et al. 1993); (2) often human miRNAs are located at fragile sites in the genome or regions that are commonly amplified or deleted in human cancer (Calin et al. 2004). For example, the deletion on chromosome 13, which is the most frequent chromosomal abnormality in chronic lymphocytic leukemia (CLL), results in the loss of two miRNA genes, *mir-15* and *mir-16,* located within this 30-kb deletion; and 3) malignant tumors and tumor cell lines have widespread dysregulated miRNA expression compared to normal tissues (Calin and Croce 2006; Gaur et al. 2007; Lu et al. 2005]. In cancer, miRNAs can be either up-regulated or down-regulated. Up-regulated miRNAs appear to target tumor suppression pathways and down-regulated miRNAs target oncogenic pathways. Among the first dysregulated miRNAs reported was miR-21, which remains the most highly dysregulated miRNA in a wide variety of cancers (Chan et al. 2005). Markedly elevated miR-21 levels occurred in human glioblastoma tumor tissues, early-passage glioblastoma cultures, and in six established glioblastoma cell lines (A172, U87, U373, LN229, LN428, and LN308) compared to non-neoplastic fetal and adult brain tissues and compared to cultured non-neoplastic glial cells. Knockdown of miR-21 in cultured glioblastoma cells triggers activation of caspases and leads to increased apoptotic cell death. A more detailed study of miR-21 targets revealed multiple important components of the p53, transforming growth factor-beta (TGF-beta), and mitochondrial apoptosis tumor-suppressive pathways (Papagiannakopoulos et al. 2008). Down-regulation of miR-21 in glioblastoma cells led to de-repression of these pathways, causing repression of growth, increased apoptosis, and cell cycle arrest. These phenotypes were dependent on two validated miR-21 targets, HNRPK and TAp63. Thus, miR-21 is an important oncogene that targets a network of p53, TGF-beta, and mitochondrial apoptosis tumor suppressor genes in glioblastoma cells.

The six miRNAs in the miR-17-92 cluster on chromosome 13 have oncogenic potential and also target pathways related to apoptosis and proliferation (Matsubara et al. 2007; Sylvestre et al. 2007). This pathway operates as follows: the transcription factor Myc induces the *mir-17-92* cluster and the expression of E2F1 growth factor. The *mir-17-92* cluster inhibits E2F1 expression (O'Donnell et al. 2005). Although up-regulation or down-regulation of miRNAs has an unclear relationship to the initiation or progression of cancer, the miRNA profile is highly predictive of cancer phenotype (Lu et al. 2005; Blenkiron et al. 2007). However, the observed correlation between the miRNA profiles and the tumor subtypes does not satisfactorily address their causative relationship. Another miRNA family implicated in cancer is *let-7*, which targets RAS. Overexpression of the RAS oncogene is common in lung cancer, and mutations in the RAS oncogene occur in approximately 15–30% of all human cancers. Therefore, let-7 has been suggested to have a tumor suppressor role in some cancers.

3 A Specialized Role for miRNAs at the Synapse

Persistent changes in synaptic strength are locally regulated by both protein degradation and synthesis; however, the coordination of these opposing limbs is poorly understood. Recently, we found that the RISC protein MOV10 was present at synapses and was rapidly degraded by the proteasome in an NMDA receptor-mediated, activity-dependent manner. Degradation of MOV10 derepressed mRNAs that were silenced within the RISC as miRNA/mRNA duplexes. This finding linked two opposite observations that have shown both protein synthesis and degradation were involved in synaptic potentiation. For example, dysregulation of proteasome-mediated degradation by mutating a specific ubiquitin ligase impaired long-term potentiation (LTP) and contextual learning (Jiang et al. 1998), the injection of a proteasome inhibitor into hippocampal CA1 resulted in impairment of some hippocampus-dependent memory tasks (Lopez-Salon et al. 2001), and proteasome blockade can diminish late phase LTP (L-LTP) induced at Schaffer collateral-CA1 synapses, like pharmacological inhibition of translation (Fonseca et al. 2006). Interestingly, co-application of proteasome blockers with translation inhibitors largely restores L-LTP, suggesting the need for a crucial balance between protein degradation and synthesis for maintenance of L-LTP (Fonseca et al. 2006). A likely target of activity-mediated degradation is Armitage. In *Drosophila,* this key RISC (RNA-induced silencing complex) factor undergoes rapid degradation in response to a specific olfactory learning paradigm (Ashraf et al. 2006). The activity-dependent loss of Armitage released translationally suppressed synaptic mRNAs such as CaMKII (Ashraf et al. 2006).

We found that MOV10, a homolog of the Drosophila DExD-box protein Armitage (Meister et al. 2005), was present at synapses and was rapidly degraded by the proteasome in an NMDA receptor-mediated, activity-dependent manner. Rat hippocampal neurons (DIV 21) were immunostained for MOV10 and synapses were identified with the post-synaptic marker, PSD95. The MOV10 pattern was punctate and extended into the dendrites. Approximately two-thirds of the synapses had detectable MOV10. Conversely, not all the MOV10 was localized to synapses; about one-fourth of the MOV10 puncta were localized to synapses. Thus, not surprisingly, the RISC has other localizations in the cell as well as the synapse. Synaptoneurosomes prepared from rat hippocampi confirmed the synapto-dendritic localization of endogenous MOV10. When photoactivatable GFP (MOV10-PA-GFP) was expressed in cultured rat hippocampal neurons, the percentage of exogenous MOV10 puncta localized to synapses increased to nearly 75%, suggesting that the exogenous MOV10 has a relative preference for synapses. When the exogenous MOV10 was co-localized with the endogenous, over two-thirds of the puncta co-localized, suggesting that the MOV10-PA-GFP was a good reporter of the endogenous synaptic pool. Ectopic expression of MOV10-PA-GFP did not alter the protein level of another key RISC component, Argonaute.

When hippocampal neurons were depolarized with KCl (60 mM, 5 min), the endogenous MOV10 was rapidly degraded within 30 min. When neurons were pretreated with the proteasome inhibitor lactacystin (10 μM, 10 min), the degradation was rescued. Comparable degradation of ectopically expressed PA-GFP tagged MOV10 also occurred upon neuronal stimulation.

To implicate more specifically the nature of the depolarization associated with MOV10 degradation, the KCl-treated cells were pretreated with the NMDA antagonist AP5 (D(-)-2-amino-5-phosphonovaleric acid) (50 μM, 50 min). AP5 pretreatment significantly prevented the KCl-induced loss of MOV10. Consistent with this observation, neuronal stimulation by NMDA alone (20 μM, 5 min) induced rapid degradation of endogenous MOV10 and the NMDA-mediated degradation was rescued by lactacystin. Thus, like its fly ortholog, MOV10 is degraded by the proteasome in an activity-dependent manner.

The degradation of MOV10 depends upon ubiquitin. We expressed GFP-fused ubiquitin mutant with all seven lysines replaced by arginine (GFP-Ub^{KO}) or GFP-fused wild type ubiquitin (GFP-Ub; Berginket al. 2006) in hippocampal neurons. GFP-Ub^{KO} has been shown to prevent ubiquitin-dependent proteasomal degradation by blocking polyubiquitination of the target protein (Berginket al. 2006). Then we depolarized neurons expressing GFP-Ub or GFP-Ub^{KO}. After depolarization, MOV10 degradation was efficiently blocked in neurons expressing the ubiquitin mutant but not wild type ubiquitin. This finding suggests that activation of the ubiquitin pathway is a frequent concomitant of depolarization. The occurrence of ubiquitinated proteins that fail to get degraded in many neurodegenerative conditions may be linked to the synaptic activation of this pathway and saturation of the system that delivers proteins to the proteasome.

To this point, the analysis of activity-induced MOV10 degradation is an ensemble analysis. Single puncta analysis is required to observe the distribution of behaviors. Rat hippocampal neurons expressing MOV10-PA-GFP were depolarized with KCl (60 mM, 5 min) compared to an untreated control. We extracted the time profiles of the loss in fluorescence intensity for ~1,500 puncta from five neurons and generated a time constant distribution drawn from individual puncta behavior. Under basal conditions, i.e., without any stimulation, the loss of fluorescence occurred with an exponential distribution that suggested that individual fluorescent loss events were independent and uncoordinated. Following KCl treatment, the loss of fluorescent signal from the puncta appeared coordinated. Under these two conditions, the mean time constant dropped from 25 min under basal conditions to 8.0 ± 2.9 min (mean $\pm$ SD) following KCl. Approximately 45% of the puncta exhibited no effective degradation and those that did degrade did so in a location-independent manner. Those puncta in which MOV10 did not degrade following activation may reside in a functionally incompetent pool due to saturation from over-expression. Because KCl depolarization affects many voltage and ligand-gated channels, similar experiments were performed with NMDA to stimulate neurons and similar results were obtained.

4   Which mRNAs are Regulated by miRNAs at the Synapse?

Like its fly ortholog, Armitage (Cook et al. 2004; Tomari et al. 2004), the loss of MOV10 relieves translational silencing. With neuronal activation, a set of mRNAs silenced by the RISC get released into an actively translating polysome pool in the synapto-dendritic compartment and elsewhere. Detection of this set of mRNAs is possible using a method we designed called translational trap. By disrupting RISC function with RNAi constructs against MOV10, the de-repressed mRNAs move to the polysomes and an increase in their levels can be detected by qPCR. This change was observed in the fractionation profile, which showed a reduced monosomal peak and an increased polysomal fraction. It was possible to identify specifically upregulated mRNAs within the polysome fraction. We selected 54 candidates based on their synaptodendritic localization, the presence of a conserved miRNA seed match in their UTR, and a context score value from Targetscan 4.0. Twenty-four percent of the tested candidate mRNAs showed a two-fold or greater enrichment in the actively translating polyribosomal fraction after RNAi against MOV10. This pool of mRNAs is likely to be regulated by miRNAs and their steady state levels are kept low due to a greater rate of degradation. mRNAs that undergo silencing but are not degraded when duplexed to a miRNA would not be detected in the assay.

Among the top changing mRNAs were two whose translation is known to be regulated by the RISC: LIMK1 and αCaMKII. Also on the list of top changing mRNAs was Lysophospholipase 1 (Lypla1). Lypla1, also known as Acyl Protein Thioesterase 1 (APT1), removes palmitate groups from proteins, including endothelial nitric-oxide synthase (Yeh et al. 1999), G_{α} and p21 (Duncan and Gilman 1998). Palmitoylation is a reversible post-translational modification that has recently been implicated in synaptic plasticity (el-Husseini Ael and Bredt 2002) and synaptic development (Kang et al. 2008; Mukai et al. 2008). The increase of Lypla1 and αCaMKII protein levels after MOV10 knock-down was confirmed by Western blot. Because Lypla1 increased after MOV10 RNAi, it is likely that its regulation was mediated by a miRNA. A strongly predicted targeting miRNA for Lypla1 is miR-138. This relationship was validated by suppression of miR-138 in hippocampal neurons and detection of an increase in Lypla1 3' UTR fused Luciferase reporter expression as well as endogenous protein level. Mutation in the seed region of the predicted miR-138 binding site in Lypla1 3'UTR also increased the luciferase reporter expression. Further validation for the localized control of miR-138 over Lypla1 translation came from double in situ hybridization combined with immunostaining. Thirty-two percent of Synapsin I-positive puncta were labeled with a miR-138 probe and Lypla1 mRNA. In separate co-localization studies, MOV10 co-localized with Lypla1 mRNA (60%).

A single puncta analysis of local dendritic protein synthesis demonstrated new protein synthesis at synaptic sites and allowed us to calculate a time constant for new protein synthesis. The αCaMKII and Lypla1 3'UTRs were fused to the photoconvertible translation reporter, Kaede, and expressed in hippocampal neurons. To interpret our results, we first studied the diffusion properties of Kaede by FRAP and found that Kaede behavior was governed by an unbinding reaction and not by free diffusion. Presumably the tetrameric nature of Kaede that leads to aggregates even at low concentration and the restricted space of the synapse further enhance aggregate formation. Consistent with these properties, Kaede had a punctate pattern when its expression levels were low and a more uniform pattern when its expression became high.

Using Kaede, we found that NMDA receptor activation can induce local protein synthesis in addition to triggering MOV10 degradation. Green Kaede fluorescence was measured over 60 min in ~1,500 individual puncta after NMDA activation (20 μM, 5 min). New protein synthesis from the αCaMKII 3'UTR-driven Kaede reporter occurred with a time constant of 11 min, i.e., after 11 min, the increasing fluorescence intensity began to plateau.

Two populations of puncta were apparent: one in which activity-dependent synthesis occurred and one in which Kaede fluorescence was unchanged. Those two populations were not regionally localized within the neuron as neighboring puncta could belong to either population. Similar protein synthesis events upon NMDA stimulation were observed with the Lypla1 3'UTR-driven Kaede reporter. The timescale and the amount of protein synthesis were in good agreement with previously reported values in the literature, with different translation reporters such as membrane anchored GFP with a 1 h half-life (Aakalu et al. 2001) or Kaede (Leung et al. 2006).

To this point the experiments point to a correlation between MOV10 degradation and local protein synthesis, but they do not prove a direct link. To show a direct link, we repeated measurements of activity-induced Lypla1 3'UTR–mediated synthesis of Kaede after RNAi knockdown of MOV10 in hippocampal neurons. A basal level of Kaede in neurons over 20 min was measured and the cells were stimulated with glutamate (20 μM, 5 min) and imaged for a further 20 min. In the presence of MOV10 RNAi, the glutamate-induced increase of Kaede fluorescence was sharply reduced. Thus localized synthesis of Lypla1 mRNA is directly linked to the integrity of MOV10. Likewise, pretreatment of neurons with lactacystin (10 μM, 10 min) blocked Lypla1 3'UTR-driven new protein synthesis from the Kaede reporter upon stimulation with glutamate (20 μM, 5 min).

5   Function of miRNA Regulation of Translation at Synapses

In addition to the depalmitoylating enzyme, Lypla1, two palmitoylating enzymes, Zdhhc2 and Zdhhc17, appeared in the trap. Palmitoylation is a widely used modification of many synaptic proteins, including Cdc42, that can potentially regulate post-synaptic structures (Kang et al. 2008). Furthermore, miR-138-mediated regulation of Lypla1 can modulate spine morphology through Rho-dependent pathways (Siegel et al. 2009). Thus, activity-dependent regulation of the synthesis of these palmitoylating and depalmitoylating enzymes represents an important facet of the mRNA repertoire under the control of the RISC. Also present in the trap was

LIMK1, a target of miR-134. The local regulation of LIMK translation also affects spine morphology (Schratt et al. 2006). Thus, one function of local translational regulation at synapses is to control changes in spine morphology, which must be coordinated through a complex set of enzymatic reactions. Interestingly, in these examples, regulation occurs over enzyme levels rather than substrates. In many reactions studied in vitro, the amount of the enzyme is not limiting; however, if regulation at the synapse affects enzyme level, one might deduce that the enzyme levels in these cases are limiting. In the case of palmitoylation and depalmitoylation, these enzymes have multiple substrates, and perhaps in the context of multiple competing substrates, the enzyme level does become limiting. Altering the level of these enzymes will likely have broad effects on many substrates by capturing substrates with higher Kms. A second outcome that arises from miRNA targeting of an enzyme is the conversion of a non-enzymatic stoichiometric reaction between the miRNA and its target mRNA to a kinetic outcome that is driven enzymatically. Thus, mass action sets the initial parameters through the miRNA binding and once depolarization the outcome is rapidly implemented according to Michaelis-Menton kinetics.

In conclusion, degradation of MOV10 and possibly other components of the RISC may underlie observations that both degradation and synthesis are required for synaptic plasticity (Ashraf et al. 2006; Fonseca et al. 2006). Interestingly, degrading MOV10 rather than recycling it places distinct kinetic boundaries on the window of translational competence at the synapse. While translation is relieved, the polysomes will be competent to synthesize a potentially computable number of copies of a protein until re-silenced by re-formation of the holo-RISC that includes MOV10 and the window of local translation closes.

Acknowledgments This work was funded by a W. M. Keck Foundation Grant and a Hillblom Foundation Grant to KSK. PN research is supported in part by the National Science Foundation under Grant No. PHY05-51164.

References

Aakalu G, Smith WB, Nguyen N, Jiang C, Schuman EM (2001) Dynamic visualization of local protein synthesis in hippocampal neurons. Neuron 30:489–502
Ashraf SI, McLoon AL, Sclarsic SM, Kunes S (2006) Synaptic protein synthesis associated with memory is regulated by the RISC pathway in Drosophila. Cell 124:191–205
Bartel DP (2004) MicroRNAs: genomics, biogenesis, mechanism, and function. Cell 116:281–297
Bergink S, Salomons FA, Hoogstraten D, Groothuis TA, de Waard H, Wu J, Yuan L, Citterio E, Houtsmuller AB, Neefjes J, Hoeijmakers JH, Vermeulen W, Dantuma NP (2006) DNA damage triggers nucleotide excision repair-dependent monoubiquitylation of histone H2A. Genes Dev 20:1343–1352
Bernstein E, Caudy AA, Hammond SM, Hannon GJ (2001) Role for a bidentate ribonuclease in the initiation step of RNA interference. Nature 409:363–366
Bernstein E, Kim SY, Carmell MA, Murchison EP, Alcorn H, Li MZ, Mills AA, Elledge SJ, Anderson KV, Hannon GJ (2003) Dicer is essential for mouse development. Nat Genet 35: 215–217

Blenkiron C, Goldstein LD, Thorne NP, Spiteri I, Chin SF, Dunning M, Barbosa-Morais NL, Teschendorff A, Green AR, Ellis IO, Tavaré S, Caldas C, Miska EA (2007) MicroRNA expression profiling of human breast cancer identifies new markers of tumour subtype. Genome Biol 8(10):R214

Brennecke J, Hipfner DR, Stark A, Russell RB, Cohen SM (2003) bantam encodes a developmentally regulated microRNA that controls cell proliferation and regulates the proapoptotic gene hid in Drosophila. Cell 113:25–36

Brennecke J, Stark A, Russell RB, Cohen SM (2005) Principles of microRNA-target recognition. PLoS Biol 3:e85

Calin GA, Croce CM (2006) MicroRNA signatures in human cancers. Nat Rev Cancer 6:857–866

Calin GA, Sevignani C, Dumitru CD, Hyslop T, Noch E, Yendamuri S, Shimizu M, Rattan S, Bullrich F, Negrini M, Croce CM (2004) Human microRNA genes are frequently located at fragile sites and genomic regions involved in cancers. Proc Natl Acad Sci USA 101:2999–3004

Chan JA, Krichevsky AM, Kosik KS (2005) MicroRNA-21 is an antiapoptotic factor in human glioblastoma cells. Cancer Res 65(14):6029–6033

Chen CZ, Li L, Lodish HF, Bartel DP (2004) MicroRNAs modulate hematopoietic lineage differentiation. Science 303:83–86

Cook HA, Koppetsch BS, Wu J, Theurkauf WE (2004) The Drosophila SDE3 homolog armitage is required for oskar mRNA silencing and embryonic axis specification. Cell 116:817–829

Duncan JA, Gilman AG (1998) A cytoplasmic acyl-protein thioesterase that removes palmitate from G protein alpha subunits and p21(RAS). J Biol Chem 273:15830–15837

el-Husseini Ael D, Bredt DS (2002) Protein palmitoylation: a regulator of neuronal development and function. Nat Rev Neurosci 3:791–802

Fonseca R, Vabulas RM, Hartl FU, Bonhoeffer T, Nagerl UV (2006) A balance of protein synthesis and proteasome-dependent degradation determines the maintenance of LTP. Neuron 52:239–245

Gaur A, Jewell DA, Liang Y, Ridzon D, Moore JH, Chen C, Ambros VR, Israel MA (2007) Characterization of microRNA expression levels and their biological correlates in human cancer cell lines. Cancer Res 67:2456–2468

Hébert SS, De Strooper B (2009) Alterations of the microRNA network cause neurodegenerative disease. Trends Neurosci 32(4):199–206

Hutvagner G, McLachlan J, Pasquinelli AE, Balint E, Tuschl T, Zamore PD (2001) A cellular function for the RNA-interference enzyme Dicer in the maturation of the let-7 small temporal RNA. Science 293:834–838

Jackson RJ, Standart N (2007) How do microRNAs regulate gene expression? Sci STKE 2007:re1

Jiang YH, Armstrong D, Albrecht U, Atkins CM, Noebels JL, Eichele G, Sweatt JD, Beaudet AL (1998) Mutation of the Angelman ubiquitin ligase in mice causes increased cytoplasmic p53 and deficits of contextual learning and long-term potentiation. Neuron 21:799–811

Kang R, Wan J, Arstikaitis P, Takahashi H, Huang K, Bailey AO, Thompson JX, Roth AF, Drisdel RC, Mastro R, Green WN, Yates JR 3rd, Davis NG, El-Husseini A (2008) Neural palmitoyl-proteomics reveals dynamic synaptic palmitoylation. Nature 456:904–909

Kosik KS (2006) The neuronal microRNA system. Nat Rev Neurosci 7(12):9

Lagos-Quintana M, Rauhut R, Lendeckel W, Tuschl T (2001) Identification of novel genes coding for small expressed RNAs. Science 294:853–858

Lau NC, Lim LP, Weinstein EG, Bartel DP (2001) An abundant class of tiny RNAs with probable regulatory roles in Caenorhabditis elegans. Science 294:858–862

Lee RC, Ambros V (2001) An extensive class of small RNAs in Caenorhabditis elegans. Science 294:862–864

Lee RC, Feinbaum RL, Ambros V (1993) The C. elegans heterochronic gene lin-4 encodes small RNAs with antisense complementarity to lin-14. Cell 75:843–854

Lee Y, Ahn C, Han J, Choi H, Kim J, Yim J, Lee J, Provost P, Radmark O, Kim S, Kim VN (2003) The nuclear RNase III Drosha initiates microRNA processing. Nature 425:415–419

Leung KM, van Horck FP, Lin AC, Allison R, Standart N, Holt CE (2006) Asymmetrical beta-actin mRNA translation in growth cones mediates attractive turning to netrin-1. Nat Neurosci 9:1247–1256

Lopez-Salon M, Alonso M, Vianna MR, Viola H, Mello e Souza T, Izquierdo I, Pasquini JM, Medina JH (2001) The ubiquitin-proteasome cascade is required for mammalian long-term memory formation. Eur J Neurosci 14:1820–1826

Lu J, Getz G, Miska EA, Alvarez-Saavedra E, Lamb J, Peck D, Sweet-Cordero A, Ebert BL, Mak RH, Ferrando AA, Downing JR, Jacks T, Horvitz HR, Golub TR (2005) MicroRNA expression profiles classify human cancers. Nature 435:834–838

Matsubara H, Takeuchi T, Nishikawa E, Yanagisawa K, Hayashita Y, Ebi H, Yamada H, Suzuki M, Nagino M, Nimura Y, Osada H, Takahashi T (2007) Apoptosis induction by antisense oligonucleotides against miR-17-5p and miR-20a in lung cancers overexpressing miR-17-92. Oncogene 26:6099–6105

Meister G, Landthaler M, Peters L, Chen PY, Urlaub H, Luhrmann R, Tuschl T (2005) Identification of novel argonaute-associated proteins. Curr Biol 15:2149–2155

Miska EA (2005) How microRNAs control cell division, differentiation and death. Curr Opin Genet Dev 15:563–568

Miska EA, Alvarez-Saavedra E, Townsend M, Yoshii A, Sestan N, Rakic P, Constantine-Paton M, Horvitz HR (2004) Microarray analysis of microRNA expression in the developing mammalian brain. Genome Biol 5:R68

Mukai J, Dhilla A, Drew LJ, Stark KL, Cao L, MacDermott AB, Karayiorgou M, Gogos JA (2008) Palmitoylation-dependent neurodevelopmental deficits in a mouse model of 22q11 microdeletion. Nat Neurosci 11:1302–1310

O'Donnell KA, Wentzel EA, Zeller KI, Dang CV, Mendell JT (2005) c-Myc-regulated microRNAs modulate E2F1 expression. Nature 435:839–843

Pillai RS, Bhattacharyya SN, Filipowicz W (2007) Repression of protein synthesis by miRNAs: how many mechanisms? Trends Cell Biol 17:118–126

Papagiannakopoulos T, Shapiro A, Kosik KS (2008) MicroRNA-21 targets a network of key tumor-suppressive pathways in glioblastoma cells. Cancer Res 68(19):8164–8172

Schratt GM, Tuebing F, Nigh EA, Kane CG, Sabatini ME, Kiebler M, Greenberg ME (2006) A brain-specific microRNA regulates dendritic spine development. Nature 439:283–289

Siegel G, Obernosterer G, Fiore R, Oehmen M, Bicker S, Christensen M, Khudayberdiev S, Leuschner PF, Busch CJ, Kane C, Hübel K, Dekker F, Hedberg C, Rengarajan B, Drepper C, Waldmann H, Kauppinen S, Greenberg ME, Draguhn A, Rehmsmeier M, Martinez J, Schratt GM (2009) A functional screen implicates microRNA-138-dependent regulation of the depalmitoylation enzyme APT1 in dendritic spine morphogenesis. Nat Cell Biol 11: 705–716

Sylvestre Y, De Guire V, Querido E, Mukhopadhyay UK, Bourdeau V, Major F, Ferbeyre G, Chartrand P (2007) An E2F/miR-20a autoregulatory feedback loop. J Biol Chem 282: 2135–2143

Tomari Y, Du T, Haley B, Schwarz DS, Bennett R, Cook HA, Koppetsch BS, Theurkauf WE, Zamore PD (2004) RISC assembly defects in the Drosophila RNAi mutant armitage. Cell 116:831–841

Yeh DC, Duncan JA, Yamashita S, Michel T (1999) Depalmitoylation of endothelial nitric-oxide synthase by acyl-protein thioesterase 1 is potentiated by Ca(2+)-calmodulin. J Biol Chem 274:33148–33154

Molecular Mechanisms for the Initiation and Maintenance of Long-Term Memory Storage

Sathyanarayanan Puthanveettil and Eric Kandel

Abstract As a result of advances in cellular, molecular and systems biology the mechanisms underlying the initial storage, maintenance and recall of memories are now beginning to be understood. To obtain these insights several model systems and methodologies have been used. Here we briefly summarize this recent progress based on study of two model systems, the gill-withdrawal reflex of *Aplysia* and hippocampal based spatial memory storage in the mouse.

How do long-term memories, once initially formed, remain stable for long periods of time ranging from weeks to months to years to the lifetime of an individual? Modern behavioral and biological studies have revealed that memory is not a unitary faculty of the mind but consists of explicit (or declarative) and implicit (or procedural) memory storage (Polster et al. 1991; Squire and Zola-Morgan 1991). Explicit memory is the conscious recall of knowledge about people, places and things and is particularly well developed in the vertebrate brain. Implicit or nondeclarative memory is memory for motor and perceptual skills. It is evolutionarily conserved and found in all creatures with a nervous system. It is expressed through performance, without conscious recall of past experience. Implicit memory includes simple associative forms of memory, such as classical and operant conditioning, and nonassociative forms, such as sensitization and habituation. Explicit and implicit memories differ not only in the information stored but also in their location within the brain (Milner 1985; Polster et al. 1991; Kandel and Squire 2000). Explicit memory is critically dependent on the hippocampal formation and structures in the medial temporal lobe of the cerebral cortex. By contrast, implicit memory represent a family of different processes that are represented in a number of brain systems, including the cerebellum, the striatum, the amygdala of vertebrates and, in the simplest cases as in invertebrates, in the sensory and motor pathways recruited during the learning process for particular perceptual or motor skills.

E. Kandel (✉)
Department of Neuroscience, Howard Hughes Medical Institute, Columbia University, New York, NY 10032, USA
e-mail: erk5@columbia.edu

We have used two experimental model systems to study representative examples of explicit and implicit memory storage: sensitization in the marine snail *Aplysia californica* as an example of implicit memory, and spatial memory formation in the rodent hippocampus as an example of explicit memory. In both model systems, the elementary events that underlie synaptic plasticity, the ability of neurons to modulate the strength of their synapses in response to extra- or intracellular cues, are thought to be fundamental for both the fine-tuning of synaptic connections during development and for behavioral learning and memory storage in the adult. Cell biological and molecular studies in both *Aplysia* and the hippocampus have provided insight into the molecular mechanism whereby activity-dependent modulation of synaptic function is encoded, processed and stored within the brain (Bliss 2003; Bliss et al. 2003; Kandel 2001).

1 Memory Has Both A Short-term and A Long-Term Component

For both implicit and explicit memory, two general types of storage mechanisms have been described: short-term memory, lasting minutes, and long-term memory, lasting days, weeks or longer. This temporal distinction in behavior is reflected in specific forms of synaptic plasticity that underlie each form of behavioral memory as well as specific molecular requirements for each of these two forms of synaptic plasticity. These two different phases differ not only in their time courses but also in their underlying molecular mechanisms (Goelet et al. 1986). Recent studies in *Aplysia* and mice have revealed that these distinct stages in behavioral memory are reflected in distinct phases of synaptic plasticity (Bailey et al. 1996; Bliss and Collingridge 1993). These different stages have been particularly well studied in the context of sensitization, a form of learning in which an animal learns to strengthen its reflex responses to previously neutral stimuli following presentation of an aversive stimulus. Neural circuitry underlying this form of learning has been delineated (Kandel 2001). The circuit is located in the abdominal ganglion and has 24 central mechanoreceptor sensory neurons that innervate the siphon skin and make direct monosynaptic connections with six gill motor cells. The sensory neurons also make indirect connections with the motor cells through small groups of excitatory and inhibitory interneurons. The sensory and motor neuron of this circuitry can be reconstituted in culture (Montarolo et al. 1986). The short and long-term memory of sensitization can be studied not only in the intact animal (Frost et al. 1985) but also in an in vitro reconstituted circuitry consisting of a sensory neuron cultured together with target motor neuron (Montarolo et al. 1986) (Fig. 1).

Here we briefly summarize recent advances in our understanding of molecular requirements for these different forms of memory storage, with a special emphasis on the molecular basis for the persistence of long-term memory storage in *Aplysia*.

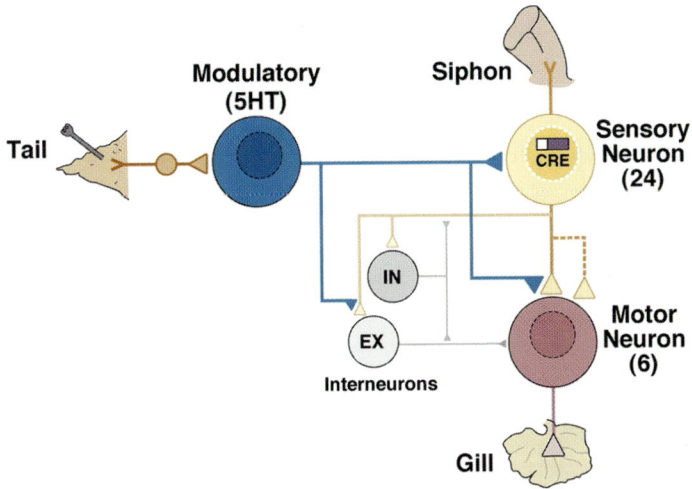

Fig. 1 Neural circuitry of the *Aplysia* gill-withdrawal reflex. The siphon is innervated by 24 sensory neurons that connect directly with the six motor neurons. The sensory neurons also connect to populations of excitatory and inhibitory interneurons that in turn connect with the motor neurons. Stimulation of the tail activates three classes of modulatory interneurons (serotonergic neurons, neurons that release the small cardioactive peptide, and the L29 cells) that act on the terminals of the sensory neurons as well as on those of the excitatory interneurons. Repeated stimulation of the tail causes CREB-mediated transcriptional activation and growth of new synaptic connections (from Kandel 2001). The sensory and motor neuron connections of this circuitry can be reconstituted in vitor (Montorolo et al. 1986). As in the intact animal, repeated stimulation of these neuronal cultures with 5HT produces long lasting changes in synaptic strength

2   Initiation: Transcription, Transport and Local Translation

In the intact animal or in the sensory to motor neuron culture system, application of a single tail shock or one pulse of serotonin (5HT), a neurotransmitter released by the interneurons activated by the sensitizing tail stimuli, produces a PKA-mediated short-term facilitation (STF) lasting several minutes and evident in an enhancement of transmitter release from the terminals of the sensory neuron. By contrast, with five spaced tail shocks or pulses of 5HT, PKA translocates into the nucleus. In doing so, it recruits MAP kinase and together they activate the transcription factor CREB1 and the removal of the repressive influence of CREB-2, resulting in activation of several immediate early genes (Alberini et al. 1994; Bartsch et al. 1995, 1998; Dash et al. 1990; Hegde et al. 1993). This cascade of events produces the long-term synaptic facilitation (LTF) lasting for more than 24 h and the growth of new synaptic connections. The simplicity of the in vitro reconstituted neuronal circuitry of the *Aplysia* gill-withdrawal reflex has allowed us to identify several key regulators of these processes (Fig. 2; Kandel 2001; Lee et al. 2007). A similar set of gene induction mediated by CREB is involved in fly and mouse (Bourtchuladze

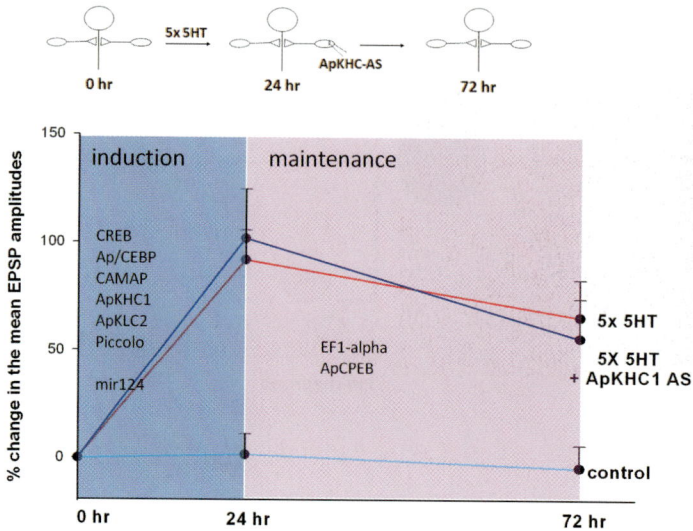

Fig. 2 Temporal phases of long-term facilitation (LTF). There are at least two temporally distinguishable phases of LTF. The first phase, described as initiation or induction, lasts up to 24 h and involves activation of gene expression and initiation of growth. During the persistence or maintenance phase, regulators of local translation play a crucial role. Examples of genes that play a major role in these two temporal phases are shown in the figure. Interestingly, inhibition of fast axonal transport by kinesin using antisense oligonucleotides (ApKHC1 AS) does not block the persistence phase

et al. 1994; Davis and Dauwalder 1991; Yin et al. 1994, 1995). LTF has three additional features: (1) kinesin-mediated, fast axonal transport in presynaptic neuron, (2) kinesin-mediated, fast axonal transport in postsynaptic neuron (Puthanveettil et al. 2008) and (3) local protein synthesis of proteins for the remodeling of pre-existing connections and formation of new connections (Martin et al. 1997).

The kinesin super family (KIF) of molecular motors transports organelles, mRNAs and protein complexes in a microtubule and ATP-dependent manner (Goldstein and Yang 2000; Hirokawa 1998; Vale and Fletterick 1997). Kinesins have important functions in chromosomal and spindle movements during mitosis and meiosis and in axonal transport. The kinesin transport machinery is composed of a heavy chain subunit (KHC) containing a conserved motor domain that attaches to the microtubule tracts and a light chain (KLC) subunit that is thought to confer unique cargo binding and regulatory specificity. The kinesin motor moves at a rate of ~1 μm per second and takes 8-nm steps from one tubulin dimer to the adjacent one, with a work efficiency of 40% (Vale and Fletterick 1997) with the application of five pulses of 5HT transcription of specific isoforms of kinesins is rapidly induced in pre- and postsynaptic neurons. This induction requires activation of cAMP-PKA-CREB pathway. Application of forskolin or microinjection of phosphoCREB into the nucleus of sensory neurons also induces mRNAs of *Aplysia* Kinesin Heavy Chain 1 (ApKHC1).

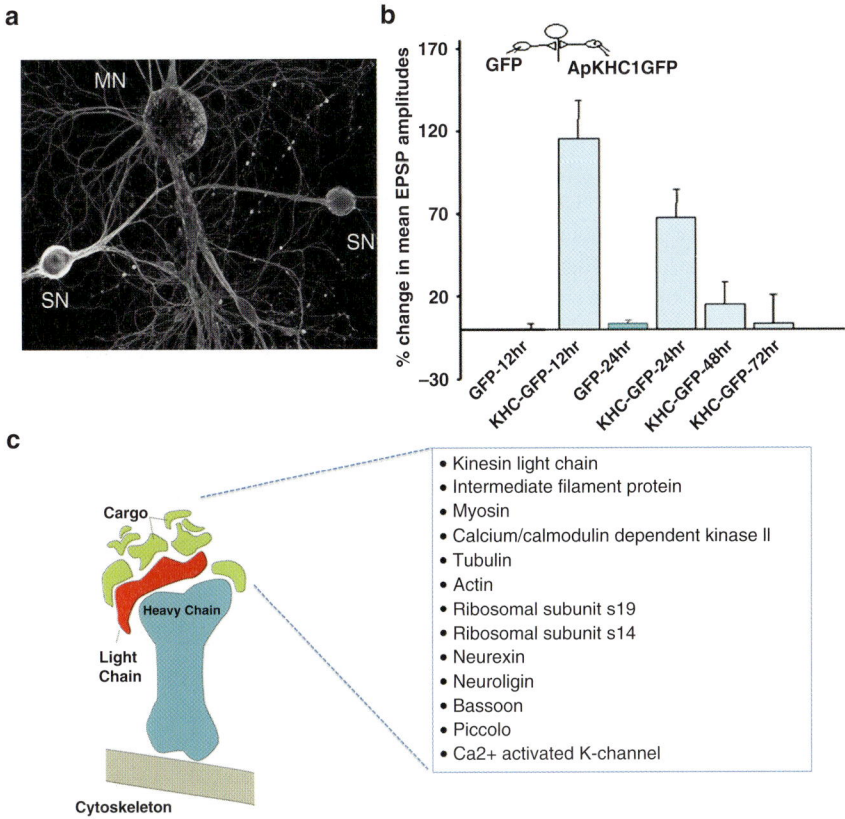

Fig. 3 Molecular motor kinesin is a key mediator for LTF. Kinesin function in pre- and postsynaptic neurons is critical for the establishment of LTF. **a**: Immunostaining of ApKHC in Aplysia sensory neuron (SN)-motor neuron (MN) cultures. Kinesin is overexpressing in the left sensory neuron. **b**: Kinesin overexpression in sensory neuron produces LTF. Excitatory post synaptic potentials (EPSPs) were measured at 12, 24 and 48 h after kinesin overexpression. **c**: Illustration of the kinesin complex. Using proteomics and biochemical methods several protein cargos were identified. Listed are the protein cargos identified from the *Aplysia* nervous system (Puthanveettil et al. 2008)

Although kinesin is required for the initial establishment of LTF, it is not required for its persistence or for STF. The microinjection of antisense oligo of ApKHC1 into sensory or motor neurons blocks the induction of LTF, and over-expression of ApKHC1 induces long-term facilitation (Fig. 3a, b). This new critical kinesin component provides a new mechanism for memory storage: kinesin transports proteins, RNAs and organelles in the early phase of memory storage so that synapses become autonomous once LTF is established. An analysis of molecules transported by kinesin revealed that molecules important for synapse formation, such as neurexin and neuroligin, were actively transported as proteins (Fig. 3c). Transported cargos also include many different RNAs (Puthanveettil et al., unpublished data) and provide a major insight into synaptic translation that is critical for the persistence of memory storage.

Not all transcriptional regulation involves immediate early genes. Some genes such as the translational regulator EF1alpha are also induced late. Ap-eEF1A is induced by 5HT as a late gene (Giustetto et al. 2003). Induction of eEF1A occurs during a second wave of transcription and translation that takes place 6–12 h after the stimulation by 5HT. Giustetto et al. (2003) found that proteins synthesized during this period are crucial for maintaining, but not for inducing, LTF. Thus late gene expression is needed for consolidating the facilitation and stabilizing new synapses. Consistent with this idea, inhibiting eEF1A blocked the late phase of LTF.

Another key distinction between the short-term and long-term processes is the regulation of the long-term process by specific microRNAs (Rajasethupathy et al. 2009). The major regulator of transcriptional control, CREB, is directly regulated by microRNA mir124. Unlike STF, establishment of LTF is regulated by ectopic regulation of mir124 (Fig. 2), which adds another level of regulation to transcriptional and translational control of memory storage.

3   Initiation: Synapse-Specific Induction of Long-Term Memory

The requirement of transcription for the memory process poses a cell-biology problem (Fig. 4a): are the long-lasting changes in neuronal plasticity induced during learning cell-wide or specific to a stimulated synapse? To address this question, Martin et al. (1997) developed a culture system in *Aplysia* in which a single, bifurcated sensory neuron of the gill-withdrawal reflex was plated in contact with two spatially separated motor neurons. In this system, application of a single pulse of 5HT to one of the two sets of synapses induced synapse-specific LTF whereas five pulses of 5HT resulted in synapse-specific LTF and structural changes (Fig. 4b). Branch-specific LTF requires local protein synthesis in the presynaptic but not the postsynaptic cell. Furthermore, presynaptic sensory neuronal processes without their cell bodies are capable of protein synthesis. This protein synthesis is stimulated three-fold by exposure to five pulses of 5HT, suggesting the existence of active and modulatable translational machinery at the synapse (Martin et al. 1997).

Interestingly, the synapse-specific LTF that is initiated in one branch of bifurcated sensory neurons can be captured at the other branch by application of a single pulse of 5HT (which by itself is capable of producing only STF; Fig. 3c). Frey and Morris and later Barco and colleagues made similar observations for the synaptic capture of long-term potentiation (LTP) in slices of mammalian hippocampus (Barco et al. 2002; Davis and Dauwalder 1991; Frey and Morris 1997). Importantly, local protein synthesis is not required for initiation of capture but is critical for the maintenance of LTF that accompanies synaptic growth (Casadio et al. 1999).

How is such a synapse-specific facilitation achieved? The capture of the synapse-specific LTF initiated in one branch by the other branch led Martin and

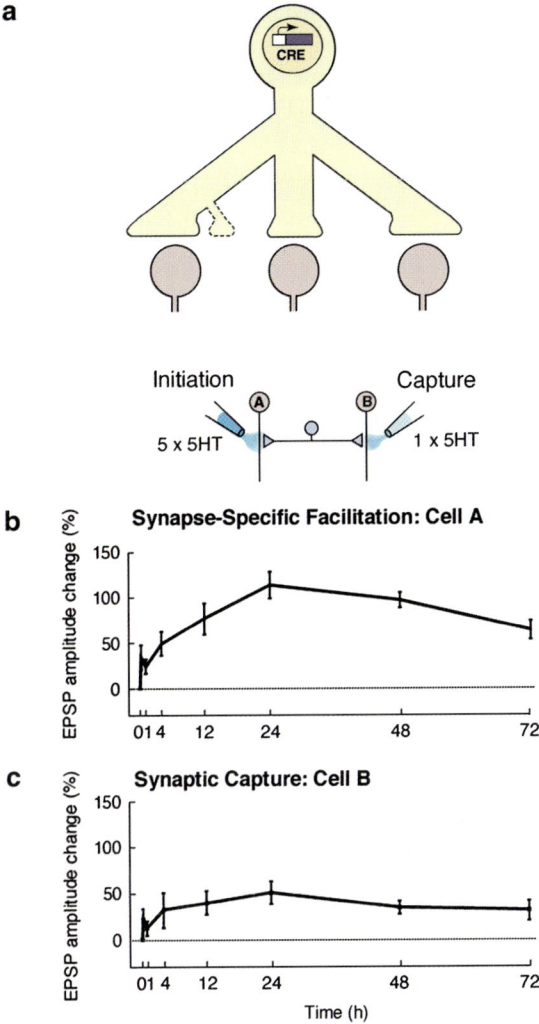

Fig. 4 Synapse-specific facilitation and synaptic capture. (**a**) Synapse-specific memory storage is a cell biological paradox. All the synapses of a neuron share a single nucleus. One would expect that the recruitment of the nucleus for the long-term process would result in cell-wide facilitation. Contrary to this notion, memory storage is synapse-specific. In this illustration, a single neuron makes connections with three different target neurons. However, only connections with one neuron are strengthened due to stimulation. How then is synapse-specific memory storage achieved? Is it possible to have delivery of molecules only to the stimulated synapses or is there a mechanism that exists at the synapse whereby only stimulated synapses can utilize the delivered molecules? (**b**) In the bifurcated sensory neuron, application of five pulses of 5HT leads to synapse-specific facilitation lasting several days. This facilitation is sensitive to disruption by transcriptional and translational inhibitors. (**c**) This synapse-specific facilitation can be captured and maintained by the other branch with the application of a single pulse of 5HT (Martin et al. 1997; Casadio et al. 1999)

coworkers to postulate that the repeated pulses of 5HT serve at least two functions: (1) they send a signal from the synapse to the nucleus that activates transcription and (2) they mark the activated synapse (Martin et al. 1997). The newly synthesized mRNAs and proteins necessary for LTF are then presumed to be transported to all the synapses of the neuron, but these gene products are only productively utilized by the marked synapse. Synaptic capture by a single pulse of serotonin further suggests that the signaling required for STF can produce the synaptic mark (Martin et al. 1997).

In exploring the nature of the synaptic mark, Casadio et al. (1999) found that it has at least two components: (1) a PKA-dependent component needed for the initial capture of synapse-specific facilitation and for the initiation of growth of new synaptic connections and (2) a rapamycin-sensitive, local protein synthesis-dependent component required for the long-term maintenance of facilitation and stabilization of growth beyond 24 h. Because mRNAs are made in the cell body, the need for the local translation of some mRNAs suggests that these mRNAs may be dormant before they reach the site of translation. If that is true, then the synaptic mark for stabilization might be a regulator of translation capable of activating translationally dormant mRNAs.

4 Persistence of Memory Storage Requires Local Cytoplasmic Polyadenylation

Interestingly, the protein synthesis inhibitor rapamycin, which blocks translation of a subset of synaptic mRNAs related to growth, does not interfere with the retrograde signal but only blocks the stabilizing component of the synaptic mark, both at the site of capture and at the site of initiation (Casadio et al. 1999). In searching for such a translational regulator, Si and colleagues (2003a) argued that, if mRNAs are going to be sent to all processes, then presumably they must be silent; otherwise, they might be activated at all synapses, whether these synapses are specifically stimulated or not. This thinking led them to focus on *c*ytoplasmic *p*olyadenylation *e*lement *b*inding protein (CPEB), a molecule that Joel Richter had discovered to activate dormant mRNAs in other biological contexts (Fig. 4a). Work on *Xenopus* oocytes revealed that some translationally dormant mRNAs were activated following elongation of their poly (A) tail (McGrew et al. 1989). This polyadenylation-dependent translational control requires two *cis*-acting elements at the 3′ UTR of the mRNAs, a polyadenylation sequence AAUAAA and a cytoplasmic polyadenylation element (CPE) with a general structure of UUUUUAU (Fox et al. 1989). Cytoplasmic polyadenylation is regulated by a CPE binding protein, CPEB (Gebauer and Richter 1996; Hake and Richter 1994). Although initially discovered in developing oocytes, CPEB was subsequently also found in cultured hippocampal neurons and in the postsynaptic density fraction of mouse synaptosomes (Huang et al. 2002; Wu et al. 1998).

Si et al. (2003a) searched for CPEB in *Aplysia* and found in addition to the developmental isoform, a new neuron specific isoform ApCPEB. The levels of this neuron specific isoform were increased by treatments with 5HT that induce LTF. Conversely local reduction of ApCPEB levels through TAT-conjugated antisense oligonucloetides blocked maintenance of LTF. Importantly, the persistence of synaptic facilitation requires ApCPEB-mediated synaptic protein synthesis not transiently but continuously, for at least 72 h (Miniaci et al. 2008).

Is CPEB a rapamycin-dependent synaptic mark? CPEB has four important features that make it an attractive candidate for a synapse-specific mark for stabilization: (1) it is activated through an extracellular signal; (2) it activates mRNAs that are translationally dormant (Stebbins-Boaz et al. 1996); (3) it is spatially restricted (Bally-Cuif et al. 1998; Schroeder et al. 1999; Tan et al. 2001) and (4) some of the mRNAs targeted by CPEB are involved in cellular growth (Chang et al. 2001; Groisman et al. 2002).

5   A Prion-Like Mechanism of CPEB Regulates Cytoplasmic Polyadenylation

The maintenance of long-term memories poses a paradoxical problem because there is constant turnover of proteins at the synapse. How does the newly formed proteins at the synapse help in remembering the previous configuration of signaling pathways at the synapse? The general solution to this problem of molecular turnover lies in a class of stable and self-sustaining biochemical reactions.

A clue for stable and self-perpetuating chemical reactions came from the analysis of amino acid sequences of ApCPEBs. The neuronal isoform of ApCPEB has a Glutamine-Asparagine (Q/ and N)-rich N-terminal domain similar to yeast prions. By contrast, the developmental isoform does not contain this Q/N-rich domain at the N-terminus. The N-terminal 160 amino acids of 44 randomly selected *Aplysia* proteins have an average glutamine + asparagine (Q + N) content of 10%, typical of proteins in other species. In contrast, the N-terminal 160 amino acids of ApCPEB have a Q + N content of 48%. A search of the protein sequence database revealed putative homologs of the *Aplysia* neuronal CPEB in *Drosophila*, mouse, and human, with N-terminal extensions of similar character. The presence of a Q/N-rich N-terminal domain of neuronal CPEB typical of yeast prion proteins led Si et al. (2003b) to test whether ApCPEB has prion-like properties (Fig. 5b, c).

The prion hypothesis suggests that biological information can be replicated exclusively through self-perpetuating conformations of proteins. Though it was initially conceived to clarify mysterious neurodegenerative diseases in mammals (Griffith 1967; Prusiner 1982), the prion-like only mechanism has since grown to encompass a number of non-Mendelian traits in fungi (Ross et al. 2005; Shkundina and Ter-Avanesyan 2007; Shorter and Lindquist 2005). All known prions, except for the initially discovered disease-causing prion, PrP, are benign

S. Puthanveettil and E. Kandel

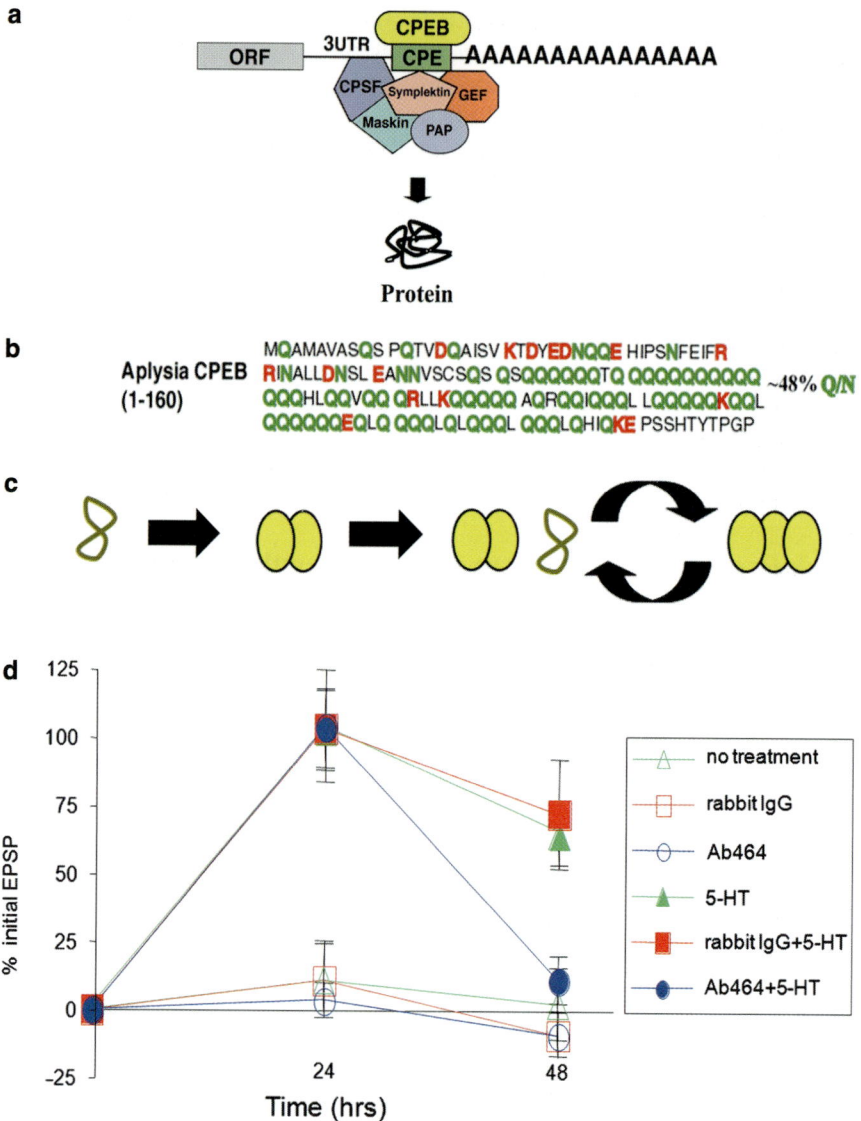

a

b

Aplysia CPEB (1-160)

MQAMAVASQS PQTVDQAISV KTDYEDNQQE HIPSNFEIFR
RINALLDNSL EANNVSCSQS QSQQQQQQTQ QQQQQQQQQQ ~48% Q/N
QQQHLQQVQQ QRLLKQQQQQ AQRQQIQQQL LQQQQQKQQL
QQQQQQEQLQ QQQLQLQQQL QQQLQHIQKE PSSHTYTPGP

c

d

Fig. 5 Neuronal isoform of *Aplysia* has prion-like properties. (**a**) CPEB is a translational regulator and binds to mRNAs containing CPE and regulates cytoplasmic polyadenylation. This polyadenylation activates translationally repressed RNAs. (**b**) *Aplysia* CPEB has a Q/N-rich N-terminal domain that is similar to the prion domain of yeast prion sup-35. (**c**) Self-perpetuation of ApCPEB through a prion-like mechanism. Application of 5HT to sensory motor neuron synapses converts CPEB in inactive conformation to active conformation. This activated CPEB further converts inactive forms to active forms, leading to self-perpetuation of local polyadenylation. (**d**) Inhibition of formation of active aggregates of ApCPEB in sensory neurons by microinjection of specific antibody (Ab464) blocks persistence of LTF (Si et al. 2010)

and, in some cases, can confer selectable advantages to the organism (Saupe 2000; True et al. 2004; True and Lindquist 2000). The self-templating property of prions makes them epigenetically dominant and enables prion-forming proteins to function as metastable cellular switches.

The realization that protein conformational switches could provide a means for inheritance of phenotypes dates back more than 15 years (Wickner 1994). However, only a few proteins with this capacity have been reported in any system (Du et al. 2008; Shorter and Lindquist 2005). Most of these proteins have been found in the yeast *S. cerevisiae,* with the [*PSI+*] element being the best understood. [*PSI+*] is caused by an amyloid-like aggregated state of the translation-termination factor Sup35p. In the prion conformation, the majority of Sup35p molecules are inactive, resulting in increased levels of nonsense suppression (Liebman and Sherman 1979; Patino et al. 1996) and programmed frame shifting (Namy et al. 2008), giving rise to RNA stability changes and functionally altered polypeptides leading to phenotypes that can be advantageous under certain conditions (Eagleston et al. 1999; True et al. 2004). In an attempt to probe for domains, Lindquist and colleagues scanned the yeast genome bioinformatically for proteins with prion-like character. They identified 24 new yeast proteins containing a prion-forming domain (PrD), thus expanding the repertoire of proteins that utilize prion-like conformational changes for its activity (Alberti et al. 2009).

The ability to undergo a conformational switch resides in structurally independent PrDs (Edskes et al. 1999; Li and Lindquist 2000; Santoso et al. 2000; Sondheimer and Lindquist 2000). These PrDs are modular and can be transferred to other proteins to confer properties of prions (Li and Lindquist 2000). They have a very unusual amino acid composition: enriched for polar residues such as glutamine (Q) and asparagine (N) and depleted of hydrophobic and charged residues. This composition promotes a disordered, molten globule-like conformational ensemble within which amyloid-nucleating contacts can be made (Mukhopadhyay et al. 2007; Serio et al. 2000; Wang et al. 2006).

In search of prion-like properties of CPEB, Si et al. (2003b) developed an assay in yeast and found that, in yeast, ApCPEB could exist in two distinct conformational states, very much like other prion proteins. One of these states is multimeric, active, and self-perpetuating (Si et al. 2003b). Based on the prion-like properties of ApCPEB in yeast, Si et al. proposed that ApCPEB exists in at least two functional states: a native state that is largely inactive, and an active state that promotes CPE-dependent polyadenylation of RNAs. Si et al. found that these two different conformation and activity levels are heritable and stable across generations in yeast. Furthermore, the N-terminal domain of the *Aplysia* neuronal CPEB can confer upon another protein, the glucocorticoid receptor, a self-perpetuating change in state with the properties of a yeast prion, implying that, similar to a prion, once in the active state ApCPEB can change the conformation of other CPEB molecules from native to active.

Si et al. (2010) then went on to explore the conformational state of ApCPEB in neurons. They found that when ApCPEB was overexpressed in sensory neurons, it formed punctate structures that are amyloid in nature, a common characteristic of all known prions. Using a fluorescence reconstitution assay in which two halves of GFP were attached to ApCPEB monomers, Si et al. showed that these punctate structures are formed by multimerization of ApCPEB monomers. Intriguingly, application of 5HT treatments that produce LTF, increased the number of detectable puncta suggesting that the multimerization is regulated by synaptic activity. Moreover, injection of an antibody that selectively binds the multimeric form of ApCPEB did not prevent the initiation of LTF but selectively blocked the maintenance (Fig. 5d). Photoconvertible GFP experiments next demonstrated that 5HT treatment can recruit new ApCPEB proteins to pre-existing multimers.

ApCPEB multimers are thus specifically involved in the long-term maintenance of synaptic plasticity. In a naive synapse, ApCPEB is thought to be monomeric and is either inactive or acts as a repressor. Synaptic activation leads to the conversion of ApCPEB to a dominant, self-sustaining, active multimeric state, which creates a self-sustaining synaptic mark limited to the specific set of synapses that have been activated. This resulted in a sustained period of translation at the activated synapse, which allows for the maintenance of the synaptic changes.

The involvement of a prion-like molecule in the formation of synapse-specific, long-term memory raises several issues. How is the autocatalytic, self-perpetuating chemical reaction resulting in synaptic enhancement regulated? One possibility is that molecular chaperones such as HSP104 ApCPEB conformational states. Indeed Si et al. (2003b) found that HSP104 is capable of reversing the active form of ApCPEB. A second consideration is that the action of ApCPEB must be highly restricted to specific synapses: if ApCPEB in the active state were to escape from a newly potentiated synapse, the entire cellular ApCPEB will be converted to active state, leading to erroneous potentiation of all synapses. This possibility suggests that there must be a mechanism whereby active CPEB is effectively restricted to the potentiated synapse. In support of this regulation, Si et al. (2010) found that the activated ApCPEB auto-associates and polymerizes, thus restricting its diffusion. These results are consistent with the idea that ApCPEB acts as a self-sustaining prion-like protein in the nervous system and thereby might allow the activity-dependent changes in synaptic efficacy to persist for long periods of time.

6   ApCPEB Represents a New Class of Functional Prions

There are several features of the prion-like state of ApCPEB that are novel, distinctly different from both the pathological PrP protein and the nonpathological but inactive state of yeast prions. First, although ApCPEB forms a large number of multimers in *Aplysia* sensory neurons, these multimers have no adverse effect either on basal synaptic functions or on survival of the neurons over days. Second, although indirect, our results suggest that, unlike other known prions that are inactive, ApCPEB retains biochemical activity in its multimeric state, such as the ability to bind mRNA. However, it is not yet evident whether these are active sites of poly(A) tail elongation or protein synthesis. Third, in all known prions, the conversion to the prion state is spontaneous. In contrast, conversion of ApCPEB to the multimeric state is regulated by a physiological signal, repeated pulses of 5HT. Finally, antibodies that recognize the multimeric form of ApCPEB selectively interfere with the persistence of synaptic facilitation beyond 24 h.

In yeast prions that have been characterized to date, the self-perpetuating prion state is biochemically inactive. The dominant self-perpetuating form of the fungal prion protein [HetS] also is associated with a genetic gain of function, although the biochemical activities associated with [HetS] are still unknown (Coustou et al. 1997). Also, a self-activating protease exhibits prion-like genetic behavior in the active state (Roberts and Wickner 2003). By contrast, the dominant self-perpetuating state of CPEB is the active state. It is able to bind CPE containing RNAs.

In addition to self-maintenance, a simple possibility is that multimerization of the ApCPEB leads to a net increase in activity by increasing the number of active units. If the monomeric form of the protein has weak activity, having multiple units would effectively increase activity. There might be other, more interesting possibilities. The prion-like multimerization requires a significant conformational conversion, which in addition to preserving the original activity might confer new activities to the protein.

These data suggest that self-perpetuating conformational reactions, reminiscent of prion conversion to the "infectious scrapie" form, might underlie the capacity of ApCPEB to self-perpetuate memory in an epigenetic fashion. The initial activation of ApCPEB begins a local positive feedback loop within the synapse induced by learning, whereby ApCPEB promotes the activation of more ApCPEB through specific conformational changes. This is potentially a very powerful mechanism for generating a self-perpetuating local signal at the synapse.

7   Conclusions

Using the advantages of the *Aplysia* sensory-motor neuron culture system, we have been able to derive a more detailed understanding of transcriptional and translational control of long-term memory storage. Several key components have been identified, including the finding that the initiation of a long-term process requires the induction of orthograde transport and that the maintenance involves a prion-like mechanism (Fig. 6). A number of key questions need to be addressed. How long does transcriptional activation last? When and how are the proteins and RNAs utilized at the synapse? What triggers the conversion of inactive CPEB to active form at the synapse? Does the conversion depend entirely on conformational changes or does it also require covalent modifications? How is the activated CPEB regulated? What are the targets of CPEB? We are currently actively pursuing these questions. Answers to these questions will have important ramifications in the biology of memory storage and will provide a greater understanding of how local protein synthesis is regulated and how long-term memories are maintained.

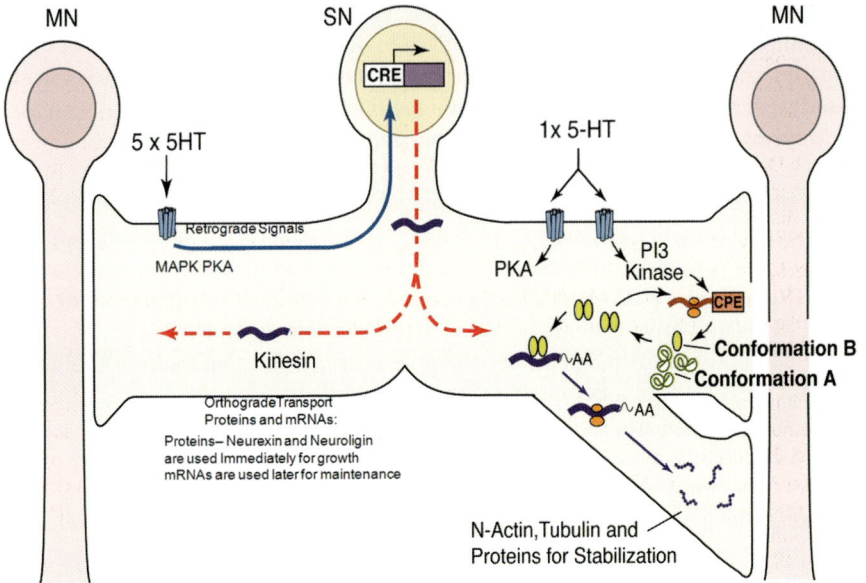

Fig. 6 A model for the initiation and persistence of long-term memory storage. A single pulse of 5HT to sensory neuron (SN) and motor neuron (MN) synapses recruits the short-term process, and this requires PKA activity. PI3 kinase also becomes activated and stimulates CPEB-dependent translation. Five pulses of 5HT activate CREB in the nucleus through the PKA-MAPK pathway and facilitate enhanced, kinesin-mediated fast axonal transport of proteins, mRNAs and organelles to synapses. These activation steps are critical for the initiation of LTF. RNAs transported by kinesin may be used for persistence. During the persistence phase, CPEB at the stimulated synapse activates polyadenylation of CPE-containing RNAs through a prion-like mechanism for the self-perpetuation of synapse-specific memory storage

In the context of a symposium on cancer and neurodegeneration, it is interesting that a variant of a prion mechanism first discovered in the context of neurodegeneration can have a functional role in the brain. It will therefore be fascinating to analyze in detail the structural and pathogenic regulatory mechanisms that constrain functional prions and prevent them from becoming pathogenic.

References

Alberini CM, Ghirardi M, Metz R, Kandel ER (1994) C/EBP is an immediate-early gene required for the consolidation of long-term facilitation in *Aplysia*. Cell 76:1099–1114

Alberti S, Halfmann R, King O, Kapila A, Lindquist S (2009) A systematic survey identifies prions and illuminates sequence features of prionogenic proteins. Cell 137:146–158

Bailey CH, Bartsch D, Kandel ER (1996) Toward a molecular definition of long-term memory storage. Proc Natl Acad Sci USA 93:13445–13452

Bally-Cuif L, Schatz WJ, Ho RK (1998) Characterization of the zebrafish Orb/CPEB-related RNA binding protein and localization of maternal components in the zebrafish oocyte. Mech Dev 77:31–47

Barco A, Alarcon JM, Kandel ER (2002) Expression of constitutively active CREB protein facilitates the late phase of long-term potentiation by enhancing synaptic capture. Cell 108:689–703

Bartsch D, Ghirardi M, Skehel PA, Karl KA, Herder SP, Chen M, Bailey CH, Kandel ER (1995) *Aplysia* CREB2 represses long-term facilitation: relief of repression converts transient facilitation into long-term functional and structural change. Cell 83:979–992

Bartsch D, Casadio A, Karl KA, Serodio P, Kandel ER (1998) CREB1 encodes a nuclear activator, a repressor, and a cytoplasmic modulator that form a regulatory unit critical for long-term facilitation. Cell 95:211–223

Bliss TV (2003) A journey from neocortex to hippocampus. Philos Trans R Soc Lond B Biol Sci 358:621–623

Bliss TV, Collingridge GL, Morris RG (2003) Introduction Long-term potentiation and structure of the issue. Philos Trans R Soc Lond B Biol Sci 358:607–611

Bliss TVP, Collingridge GL (1993) A synaptic model of memory – long-term potentiation in the hippocampus. Nature 361:31–39

Bourtchuladze R, Frenguelli B, Blendy J, Cioffi D, Schutz G, Silva AJ (1994) Deficient long-term memory in mice with a targeted mutation of the cAMP-responsive element-binding protein. Cell 79:59–68

Casadio A, Martin KC, Giustetto M, Zhu H, Chen M, Bartsch D, Bailey CH, Kandel ER (1999) A transient, neuron-wide form of CREB-mediated long-term facilitation can be stabilized at specific synapses by local protein synthesis. Cell 99:221–237

Chang JS, Tan L, Wolf MR, Schedl P (2001) Functioning of the Drosophila orb gene in gurken mRNA localization and translation. Development 128:3169–3177

Coustou V, Deleu C, Saupe S, Begueret J (1997) The protein product of the het-s heterokaryon incompatibility gene of the fungus Podospora anserina behaves as a prion analog. Proc Natl Acad Sci USA 94:9773–9778

Dash PK, Hochner B, Kandel ER (1990) Injection of the cAMP-responsive element into the nucleus of *Aplysia* sensory neurons blocks long-term facilitation. Nature 345:718–721

Davis RL, Dauwalder B (1991) The Drosophila dunce locus: learning and memory genes in the fly. Trends Genet 7:224–229

Du Z, Park KW, Yu H, Fan Q, Li L (2008) Newly identified prion linked to the chromatin-remodeling factor Swi1 in Saccharomyces cerevisiae. Nat Genet 40:460–465

Eagleston SS, Cox BS, Tuite MF (1999) Translation termination efficiency can be regulated in Saccharomyces cerevisiae by environmental stress through a prion-mediated mechanism. EMBO J 18:1974–1981

Edskes HK, Gray VT, Wickner RB (1999) The [URE3] prion is an aggregated form of Ure2p that can be cured by overexpression of Ure2p fragments. Proc Natl Acad Sci USA 96:1498–1503

Fox CA, Sheets MD, Wickens MP (1989) Poly(A) addition during maturation of frog oocytes: distinct nuclear and cytoplasmic activities and regulation by the sequence UUUUUAU. Genes Dev 3:2151–2162

Frey U, Morris RG (1997) Synaptic tagging and long-term potentiation. Nature 385:533–536

Frost WN, Castellucci VF, Hawkins RD, Kandel ER (1985) Monosynaptic connections made by the sensory neurons of the gill- and siphon-withdrawal reflex in *Aplysia* participate in the storage of long-term memory for sensitization. Proc Natl Acad Sci USA 82:8266–8269

Gebauer F, Richter JD (1996) Mouse cytoplasmic polyadenylation element binding protein: an evolutionarily conserved protein that interacts with the cytoplasmic polyadenylylation elements of c-mos mRNA. Proc Natl Acad Sci USA 93:14602–14607

Giustetto M, Hegde AN, Si K, Casadio A, Inokuchi K, Pei W, Kandel ER, Schwartz JH (2003) Axonal transport of eukaryotic translation elongation factor 1alpha mRNA couples transcription in the nucleus to long-term facilitation at the synapse. Proc Natl Acad Sci USA 100:13680–13685

Goelet P, Castellucci VF, Schacher S, Kandel ER (1986) The long and the short of long-term memory – a molecular framework. Nature 322:419–422

Goldstein LS, Yang Z (2000) Microtubule-based transport systems in neurons: the roles of kinesins and dyneins. Annu Rev Neurosci 23:39–71

Griffith JS (1967) Self-replication and scrapie. Nature 215:1043–1044

Groisman I, Jung MY, Sarkissian M, Cao Q, Richter JD (2002) Translational control of the embryonic cell cycle. Cell 109:473–483

Hake LE, Richter JD (1994) CPEB is a specificity factor that mediates cytoplasmic polyadenylation during Xenopus oocyte maturation. Cell 79:617–627

Hegde AN, Goldberg AL, Schwartz JH (1993) Regulatory subunits of cAMP-dependent protein kinases are degraded after conjugation to ubiquitin: a molecular mechanism underlying long-term synaptic plasticity. Proc Natl Acad Sci USA 90:7436–7440

Hirokawa N (1998) Kinesin and dynein superfamily proteins and the mechanism of organelle transport. Science 279:519–526

Huang YS, Jung MY, Sarkissian M, Richter JD (2002) N-methyl-D-aspartate receptor signaling results in Aurora kinase-catalyzed CPEB phosphorylation and alpha CaMKII mRNA polyadenylation at synapses. EMBO J 21:2139–2148

Kandel ER (2001) The molecular biology of memory storage: a dialogue between genes and synapses. Science 294:1030–1038

Kandel ER, Squire LR (2000) Neuroscience: breaking down scientific barriers to the study of brain and mind. Science 290:1113–1120

Lee SH, Lim CS, Park H, Lee JA, Han JH, Kim H, Cheang YH, Lee YS, Ko HG, Jang DH, Kim H, Miniaci MC, Bartsch D, Kim E, Bailey CH, Kandel ER, Kaang BK (2007) Nuclear translocation of CAM-associated protein activates transcription for long-term facilitation in *Aplysia*. Cell 129:801–812

Li L, Lindquist S (2000) Creating a protein-based element of inheritance. Science 287:661–664

Liebman SW, Sherman F (1979) Extrachromosomal psi+ determinant suppresses nonsense mutations in yeast. J Bacteriol 139:1068–1071

Martin KC, Casadio A, Zhu H, Yaping E, Rose JC, Chen M, Bailey CH, Kandel ER (1997) Synapse-specific, long-term facilitation of aplysia sensory to motor synapses: a function for local protein synthesis in memory storage. Cell 91:927–938

McGrew LL, Dworkin-Rastl E, Dworkin MB, Richter JD (1989) Poly(A) elongation during Xenopus oocyte maturation is required for translational recruitment and is mediated by a short sequence element. Genes Dev 3:803–815

Milner B, Petrides M, Smith ML (1985) Frontal lobes and the temporal organization of memory. Hum Neurobiol 4:137–142

Miniaci MC, Kim JH, Puthanveettil SV, Si K, Zhu H, Kandel ER, Bailey CH (2008) Sustained CPEB-dependent local protein synthesis is required to stabilize synaptic growth for persistence of long-term facilitation in *Aplysia*. Neuron 59:1024–1036

Montarolo PG, Goelet P, Castellucci VF, Morgan J, Kandel ER, Schacher S (1986) A critical period for macromolecular synthesis in long-term heterosynaptic facilitation in *Aplysia*. Science 234:1249–1254

Mukhopadhyay S, Krishnan R, Lemke EA, Lindquist S, Deniz AA (2007) A natively unfolded yeast prion monomer adopts an ensemble of collapsed and rapidly fluctuating structures. Proc Natl Acad Sci USA 104:2649–2654

Namy O, Galopier A, Martini C, Matsufuj S, Fabret C, Rousset JP (2008) Epigenetic control of polyamines by the prion [PSI(+)]. Nat Cell Biol 10:1069–1075

Patino MM, Liu JJ, Glover JR, Lindquist S (1996) Support for the prion hypothesis for inheritance of a phenotypic trait in yeast. Science 273:622–626

Polster MR, Nadel L, Schacter DL (1991) Cognitive neuroscience analyses of memory – a historical perspective. J Cogn Neurosci 3:95–116

Prusiner SB (1982) Novel proteinaceous infectious particles cause scrapie. Science 216:136–144

Puthanveettil SV, Monje FJ, Miniaci MC, Choi YB, Karl KA, Khandros E, Gawinowicz MA, Sheetz MP, Kandel ER (2008) A new component in synaptic plasticity: upregulation of kinesin in the neurons of the gill-withdrawal reflex. Cell 135:960–973

Rajasethupathy P, Fiumara F, Sheridan R, Betel D, Puthanveettil SV, Russo JJ, Sander C, Tuschl T, Kandel E (2009) Characterization of small RNAs in aplysia reveals a role for miR-124 in constraining synaptic plasticity through CREB. Neuron 63:803–817

Roberts BT, Wickner RB (2003) Heritable activity: a prion that propagates by covalent auto-activation. Genes Dev 17:2083–2087

Ross ED, Edskes HK, Terry MJ, Wickner RB (2005) Primary sequence independence for prion formation. Proc Natl Acad Sci USA 102:12825–12830

Santoso A, Chien P, Osherovich LZ, Weissman JS (2000) Molecular basis of a yeast prion species barrier. Cell 100:277–288

Saupe SJ (2000) Molecular genetics of heterokaryon incompatibility in filamentous ascomycetes. Microbiol Mol Biol Rev 64:489–502

Schroeder KE, Condic ML, Eisenberg LM, Yost HJ (1999) Spatially regulated translation in embryos: asymmetric expression of maternal Wnt-11 along the dorsal-ventral axis in Xenopus. Dev Biol 214:288–297

Serio TR, Cashikar AG, Kowal AS, Sawicki GJ, Moslehi JJ, Serpell L, Arnsdorf MF, Lindquist SL (2000) Nucleated conformational conversion and the replication of conformational information by a prion determinant. Science 289:1317–1321

Shkundina IS, Ter-Avanesyan MD (2007) Prions. Biochemistry (Mosc) 72:1519–1536

Shorter J, Lindquist S (2005) Prions as adaptive conduits of memory and inheritance. Nat Rev Genet 6:435–450

Si K, Giustetto M, Etkin A, Hsu R, Janisiewicz AM, Miniaci MC, Kim JH, Zhu H, Kandel ER (2003a) A neuronal isoform of CPEB regulates local protein synthesis and stabilizes synapse-specific long-term facilitation in aplysia. Cell 115:893–904

Si K, Lindquist S, Kandel ER (2003b) A neuronal isoform of the aplysia CPEB has prion-like properties. Cell 115:879–891

Si K, Choi YB, White-Grindley E, Majumdar A, Kandel ER (2010) *Aplysia* CPEB can form prion-like multimers in sensory neurons that contribute to long-term facilitation. Cell 140:421–435

Sondheimer N, Lindquist S (2000) Rnq1: an epigenetic modifier of protein function in yeast. Mol Cell 5:163–172

Squire LR, Zola-Morgan S (1991) The medial temporal lobe memory system. Science 253:1380–1386

Stebbins-Boaz B, Hake LE, Richter JD (1996) CPEB controls the cytoplasmic polyadenylation of cyclin, Cdk2 and c-mos mRNAs and is necessary for oocyte maturation in Xenopus. EMBO J 15:2582–2592

Tan L, Chang JS, Costa A, Schedl P (2001) An autoregulatory feedback loop directs the localized expression of the Drosophila CPEB protein Orb in the developing oocyte. Development 128:1159–1169

True HL, Lindquist SL (2000) A yeast prion provides a mechanism for genetic variation and phenotypic diversity. Nature 407:477–483

True HL, Berlin I, Lindquist SL (2004) Epigenetic regulation of translation reveals hidden genetic variation to produce complex traits. Nature 431:184–187

Vale RD, Fletterick RJ (1997) The design plan of kinesin motors. Annu Rev Cell Dev Biol 13:745–777

Wang X, Wang F, Arterburn L, Wollmann R, Ma J (2006) The interaction between cytoplasmic prion protein and the hydrophobic lipid core of membrane correlates with neurotoxicity. J Biol Chem 281:13559–13565

Wickner RB (1994) [URE3] as an altered URE2 protein: evidence for a prion analog in Saccharomyces cerevisiae. Science 264:566–569

Wu L, Wells D, Tay J, Mendis D, Abbott MA, Barnitt A, Quinlan E, Heynen A, Fallon JR, Richter JD (1998) CPEB-mediated cytoplasmic polyadenylation and the regulation of experience-dependent translation of alpha-CaMKII mRNA at synapses. Neuron 21:1129–1139

Yin JC, Wallach JS, Del Vecchio M, Wilder EL, Zhou H, Quinn WG, Tully T (1994) Induction of a dominant negative CREB transgene specifically blocks long-term memory in Drosophila. Cell 79:49–58

Yin JC, Del Vecchio M, Zhou H, Tully T (1995) CREB as a memory modulator: induced expression of a dCREB2 activator isoform enhances long-term memory in Drosophila. Cell 81:107–115

Index

T. Curran and Y. Christen (eds.), *Two Faces of Evil: Cancer and Neurodegeneration*,   161
Research and Perspectives in Alzheimer's Disease, DOI 10.1007/978-3-642-16602-0,
© Springer-Verlag Berlin Heidelberg 2011

LIST OF PREVIOUSLY PUBLISHED VOLUMES IN THE SERIES
RESEARCH AND PERSPECTIVES IN ALZHEIMER'S DISEASE

Printing and Binding: Stürtz GmbH, Würzburg